IRMA Lectures in Mathematics and Theoretical Physics 4

Edited by Vladimir G. Turaev

Institut de Recherche Mathématique Avancée
Université Louis Pasteur et CNRS
7 rue René Descartes
67084 Strasbourg Cedex
France

IRMA Lectures in Mathematics and Theoretical Physics

Three Courses on Partial Differential Equations

Editor

Eric Sonnendrücker

W
DE
G

Walter de Gruyter · Berlin · New York

Editor

Eric Sonnendrücker
Département de Mathématique, Université Louis Pasteur,
7, rue René Descartes, 67084 Strasbourg Cedex, France, e-mail: sonnen@math.u-strasbg.fr

Series Editor

Vladimir G. Turaev
Institut de Recherche Mathématique Avancée (IRMA), Université Louis Pasteur − C.N.R.S.,
7, rue René Descartes, 67084 Strasbourg Cedex, France, e-mail: turaev@math.u-strasbg.fr

Mathematics Subject Classification 2000:
35-02; 35R30, 35R60, 74B99

Key words:
Partial differential equations, elasticity, waves, random media, parameter identification, inverse
problems

Library of Congress Cataloging-in-Publication Data

Three courses on partial differential equations / editor, Eric
Sonnendrücker.
    p.   cm. − (IRMA lectures in mathematics and theo-
retical physics ; 4)
    Includes bibliographical references.
    ISBN 3-11-017958-X (alk. paper)
    1. Differential equations, Partial.   I. Sonnendrücker,
Eric.   II. Series.
QA377.T526   2003
515'.353−dc22

                                                        2003062470

ISBN 3-11-017958-X

Bibliographic information published by Die Deutsche Bibliothek

Die Deutsche Bibliothek lists this publication in the Deutsche Nationalbibliografie;
detailed bibliographic data is available in the Internet at <http://dnb.ddb.de>.

Preface

Partial differential equations (PDEs) have been playing a major role in the modeling of many phenomena in physics and engineering for a long time. The realm of applications is growing fast, in particular in economics and life sciences. Numerical simulation of the phenomena that are investigated with the aim of predicting their behavior is thus linked to a good understanding of the mathematical properties of the PDEs involved and the derivation of efficient and robust numerical methods to solve them. Therefore a good training in partial differential equations is essential for mathematicians interested in such applications.

Following traditional introductory graduate courses on different topics of partial differential equations ranging from theoretical to numerical aspects that were given during the year 2001 (Diplôme d'Études Approfondies de mathématiques) at the University Louis Pasteur in Strasbourg, a special week consisting of five advanced lectures was proposed to the students. These lectures were the following.

- Michel Chipot (University of Zürich): On some questions in elasticity.

- Patrick Ciarlet (ENSTA, Paris): The finite element method in non-smooth computational domains.

- Josselin Garnier (University Paul Sabatier, Toulouse): Waves in random media.

- Otared Kavian (University of Versailles Saint-Quentin): Introduction to parameter identification.

- François Murat (University Pierre et Marie Curie, Paris): Homogenisation.

This volume contains enhanced versions of the lectures of M. Chipot, J. Garnier and O. Kavian.

These three lectures all address very important and interesting, but very different aspects motivated by several applications in partial differential equations. Michel Chipot investigates equilibrium positions of several disks rolling on a wire. He is looking in particular for existence of, uniqueness of, and the exact position for an equilibrium. Josselin Garnier considers problems arising from acoustics and geophysics where waves propagate in complicated media, the properties of which can only be described statistically. He shows, in particular, that when the different scales presented in the problem can be separated, there exists a deterministic result. Otared Kavian is interested in so-called inverse problems, where one or several parameters of a PDE need to be determined, by, using for example, measurements on the boundary of the domain. The question that arises naturally is what information is necessary to determine the unknown parameters. This lecture answers to this questions in different settings.

To conclude, I would like to thank all the people who made this special week on partial differential equations possible and very interesting: Olivier Debarre who suggested that I organize it, the five lecturers, Michel Chipot, Patrick Ciarlet, Josselin Garnier, Otared Kavian, and François Murat, who all gave very clear and interesting lectures. And, of course, thanks to all the students who came from the University of Strasbourg and from several other universities in France to attend these lectures. Many thanks also to those who made this volume possible, first of all the three contributors, but also Vladimir Turaev, editor of this series, the referees and the staff at Walter de Gruyter.

Strasbourg, September 10, 2003   *E. Sonnendrücker*

Table of Contents

Elastic deformations
of membranes and wires

Michel Chipot

Institut für Mathematik, Angewandte Mathematik
Winterthurerstrasse 190, 8057 Zürich, Switzerland
email: chipot@amath.unizh.ch

Abstract. When we let roll freely balls on an elastic membrane they will stop after some time in a position that we will call equilibrium position. We would like to determine such equilibrium and see if it is unique. Since this three dimensional problem is not so easy to handle we look also to the two dimensional analogue of disks rolling on a wire.

2000 Mathematics Subject Classification: 35J85, 49J40, 49J45

1 Introduction

Consider a rigid planar frame Γ. On this frame suppose that is spanned an elastic membrane. When we drop a ball on the membrane, this ball will dig its nest in the elastic film and roll for some time before to reach some equilibrium position (see Figure 1). The same phenomenon will occur if we let roll several balls on the membrane. This is this equilibrium position that we would like to investigate from the point of view of existence, uniqueness, location.

In the case of one single ball the problem was studied in [4]. The symmetry of the heavy ball considered played a crucial rôle there. When one passes to a system formed by two different balls symmetry is lost and results are getting more challenging. Very few are in fact available and this is the reason why we are lead to consider the deformation of an elastic wire by one or several disks and more generally by two dimensional rigid bodies – like for instance a square. Then direct computations can be performed. As we will see some unexpected situations might then occur that could help to understand the three dimensional problem. For instance, in the case of two identical disks sitting on a wire, the equilibrium position reached is not always the obvious one of the two disks hanging at the same level in the middle of the wire. Tension could force them to tilt as shown on Figure 2 below when their weight become too high – see [3]. A similar loss of symmetry can occur in the case of a square for instance – see [2].

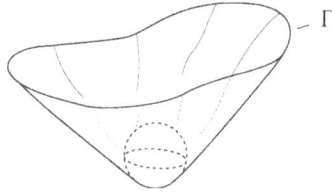

Figure 1. The case of a membrane

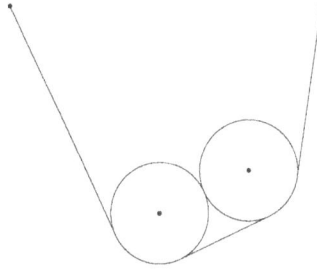

Figure 2. The case of two heavy disks on a wire

We will follow the following steps. First we will consider the case of a two dimensional object supported by a wire. Addressing first the case of a disk we will give detailed computations which will help to understand the main ingredients necessary to address the problem of two disks. Then we will consider the equilibrium of a square able to move freely on a wire. In the next section we will attack the question of two different disks. Finally we will treat the case of one ball rolling on an elastic membrane, then the case of several. All along we will point out the many open problems that are related to the different cases.

2 The case of a single two dimensional solid on a wire

2.1 The case of one disk on an elastic wire

Let us denote by Ω the interval $(0, 1)$. Ω is the undeformed configuration of a one dimensional elastic wire spanned between the two points 0 and 1 of the real axis. We will denote by $0x$ the horizontal real axis and by $0h$ the vertical one that we will suppose oriented downward (see Figure 3). Letting roll a heavy disk on this wire spanned between 0 and 1 we would like to determine its equilibrium position. Of course – for symmetry reasons – we are expecting that the disk will stabilize in the middle of the wire at a level such that the tension of the elastic support balances the weight of the disk. (We suppose an ideal situation where the two dimensional disk can be maintained vertically on the wire). However, we need yet to introduce the right energy considerations allowing us to show that.

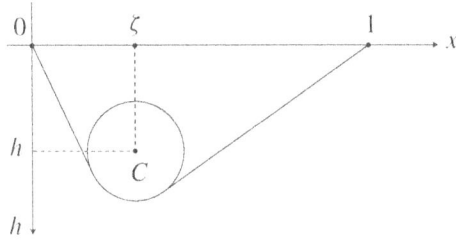

Figure 3. One single disk on a wire

Let G be the weight of the disk and r its radius. In order to allow the disk to fit entirely between the points 0, 1 we will assume

$$r < \frac{1}{2}. \qquad (2.1)$$

Then, clearly, the position of the disk in space is completely determined by the position of its center

$$C = (\zeta, h) \in \Omega \times \mathbb{R}. \qquad (2.2)$$

C being fixed the lower points of the disk can be described by the function

$$\psi^{\zeta h}(x) = \psi(x) = h + \sqrt{r^2 - (x - \zeta)^2} \quad \forall x \in B(\zeta, r), \qquad (2.3)$$

where $B(\zeta, r) = (\zeta - r, \zeta + r)$ denotes the open ball of center ζ and radius r – see the figure below. An admissible deformation of the wire is a "smooth" function u such

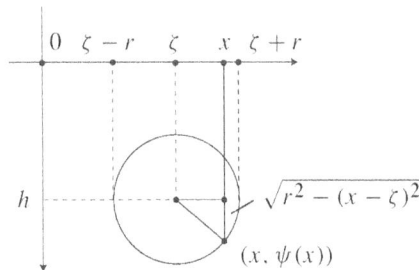

Figure 4. The function ψ

that

$$u(0) = u(1) = 0, \qquad u(x) \geq \psi(x) \quad \text{for } x \in B(\zeta, r), \tag{2.4}$$

i.e. a function u describing a reasonable state for the wire – that has to be attached at the end points of the interval $(0, 1)$ and has to be below the disk. (This is the rôle of the inequality (2.4) – recall that the vertical axis is directed downward). Now, in order to develop analytical arguments we have to choose u in some appropriate space. We will suppose

$$u \in H_0^1(\Omega) \tag{2.5}$$

where

$$H_0^1(\Omega) = \{ v \in L^2(\Omega) \mid v' \in L^2(\Omega), v(0) = v(1) = 0 \}. \tag{2.6}$$

(We refer to [6], [13], [17] for an introduction to Sobolev spaces.) Note that in our analysis we consider as "admissible", deformations for which the wire can be loose – i.e. not spanned or not being tight as in Figure 3 for example. In a case of a nonhomogeneous wire the equilibrium position of the wire could be such that the disk is not supported by straight lines as one can see using the computation that we are going to develop now. Let us suppose the wire to be homogeneous. Then, the center of the disk being located at $C = (\zeta, h)$, the energy of an admissible deformation – i.e. a function u such that

$$u \in H_0^1(\Omega), \qquad u(x) \geq \psi^{\zeta h}(x) \quad \text{a.e. } x \in B(\zeta, r) \tag{2.7}$$

is given by – if u_x denotes the derivative of u in x –

$$E(u; \zeta, h) = \frac{1}{2}\sigma \int_\Omega u_x^2(x)\,dx - Gh. \tag{2.8}$$

The integral part of this energy corresponds to the elastic energy of the system, the other part is a potential energy, σ is some constant of elasticity characteristic of the material. Nature is careful and tends to adopt behaviors that minimize energy. In other words among all possible admissible deformations the one corresponding to the equilibrium of the disk on the wire will be the one minimizing $E(u; \zeta, h)$. So we have to find an eventual triple (u_0, ζ_0, h_0) such that

$$\begin{cases} (\zeta_0, h_0) \in (r, 1-r) \times \mathbb{R}, & u_0 \text{ admissible (i.e. satisfying (2.7))}, \\ E(u_0; \zeta_0, h_0) \leq E(u; \zeta, h) & \forall (\zeta, h) \in (r, 1-r) \times \mathbb{R}, \forall u \text{ admissible}. \end{cases} \tag{2.9}$$

(We choose $\zeta \in (r, 1-r)$ in such a way that the disk cannot touch the end points 0 and 1).

Remark 2.1. The problem (2.9) is an unusual problem of the calculus of variations. The class in which we are looking for a minimizer is neither convex nor compact.

One could consider a nonhomogeneous elastic wire. In this case the energy to minimize would be

$$E(u; \zeta, h) = \frac{1}{2} \int_\Omega \sigma(x) u_x^2(x) \, dx - Gh \qquad (2.10)$$

where σ is some coefficient of elasticity. Similarly we could assume the disk to be nonhomogeneous and have a potential energy of the type

$$- \int_D G(x, y) y \, dx \, dy \qquad (2.11)$$

where the integration is taking place on the disk D of center (ζ, h) and radius r. The notation for $G(x, y)$ – the density of weight – is clear. The analysis that we develop below is widely open in the cases (2.10), (2.11). Note that in the case of (2.11) the center of mass or the position of the disk around its center are additional parameters of the problem. We leave that to the curiosity of the reader.

Let us now consider the problem (2.9). Dividing the energy (2.8) by σ it is clear that at the expense of rescaling G we can always assume that

$$\sigma = 1 \qquad (2.12)$$

This is what we will do from now on.

Somehow – even if this is wrong in the nonhomogeneous case – our intuition tells us that the wire should be tight i.e. adopt a position made by straight lines outside of the contact zone of the disk. This is what we are about to see now. For that notice that if (u_0, ζ_0, h_0) is a solution to (2.9) then it holds that:

$$E(u_0; \zeta_0, h_0) \leq E(u; \zeta_0, h_0) \qquad \forall u \text{ admissible} \qquad (2.13)$$

$$\Longleftrightarrow \quad \frac{1}{2} \int_0 u_{0x}^2(x) \, dx \leq \frac{1}{2} \int_\Omega u_x^2(x) \, dx \qquad \forall u \text{ admissible.} \qquad (2.14)$$

To describe the set of admissible deformations, for $C = (\zeta, h)$ fixed let us introduce

$$K(\zeta, h) = \{ u \in H_0^1(\Omega) \mid u(x) \geq \psi^{\zeta h}(x) \text{ a.e. } x \in B(\zeta, r) \}. \qquad (2.15)$$

Then, following (2.14), u_0 is solution of a minimization problem that we make more precise in the proposition below:

Proposition 2.1. *Let $(\zeta, h) \in (r, 1 - r) \times \mathbb{R}$ be fixed. Then, there exists a unique $u = u(\zeta, h)$ solution to*

$$\begin{cases} u \in K(\zeta, h), \\ \dfrac{1}{2} \int_\Omega u_x^2(x) \, dx \leq \dfrac{1}{2} \int_\Omega v_x^2(x) \, dx \quad \forall v \in K(\zeta, h). \end{cases} \qquad (2.16)$$

Moreover, u is the solution of the so called Variational Inequality

$$\begin{cases} u \in K(\zeta, h), \\ \displaystyle\int_\Omega u_x(v_x - u_x)\,dx \geq 0 \quad \forall v \in K(\zeta, h). \end{cases} \tag{2.17}$$

Proof. Let us denote by $|\cdot|_{1,2}$ the quantity defined as

$$|u|_{1,2} = \left\{ \int_\Omega u_x^2(x)\,dx \right\}^{1/2}. \tag{2.18}$$

It is well known – see [9], [18] – that this defines a Hilbert norm on $H_0^1(\Omega)$ which is a Hilbert space when equipped with it. We now leave to the reader to show that

$$K(\zeta, h) = K \text{ is a closed convex subset of } H_0^1(\Omega). \tag{2.19}$$

Then, (2.16) can be also written as

$$u \in K(\zeta, h), \qquad |u - 0|_{1,2} \leq |v - 0|_{1,2} \quad \forall v \in K(\zeta, h), \tag{2.20}$$

i.e. u is the projection of 0 on the closed convex set K. It is well known see for in-

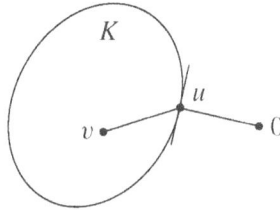

Figure 5. The projection of 0 on a closed convex set

stance [8] – that a solution u to (2.20) exists and is unique. Moreover, u is equivalently solution to – see Figure 5

$$u \in K, \qquad (0 - u, v - u)_{1,2} \leq 0 \quad \forall v \in K, \tag{2.21}$$

where $(\cdot, \cdot)_{1,2}$ denotes the scalar product on $H_0^1(\Omega)$ associated to (2.18) and defined by

$$(u, v)_{1,2} = \int_\Omega u_x v_x \, dx. \tag{2.22}$$

It is then clear that (2.21) is nothing but (2.17). This completes the proof of the proposition. $\qquad \square$

We can now describe exactly what is the solution to (2.16) or (2.17). We have:

Proposition 2.2. *For* $(\zeta, h) \in (r, 1 - r) \times \mathbb{R}$ *the solution u of* (2.16) *or* (2.17) *is given by the following function:*

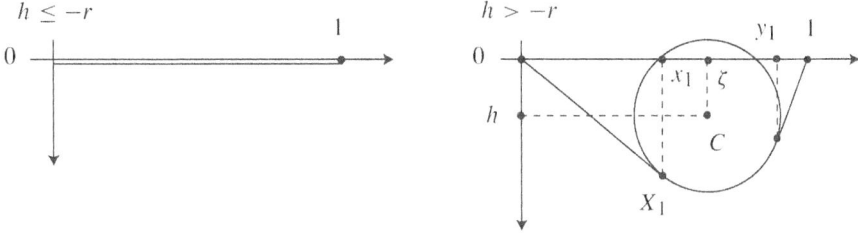

Figure 6. The solution u

$$
\begin{aligned}
i.e. \quad & if \ h \leq -r \quad && u \equiv 0, && (2.23) \\
& if \ h > -r \quad && u(x) = x \psi_x(x_1) && for \ x \in [0, x_1], \\
& && = \psi(x) && for \ x \in [x_1, y_1], && (2.24) \\
& && = (x - 1) \psi_x(y_1) && for \ x \in [y_1, 1],
\end{aligned}
$$

where ψ_x denotes the derivative of the function $\psi = \psi^{\zeta h}$ and x_1, y_1 the coordinates of the points of contact of the tangents to the disk issued from 0 and 1 – see Figure 6.

Proof. Suppose that $h \leq -r$. Then, by (2.3)

$$\psi^{\zeta h}(x) \leq 0$$

and $u = 0$ belongs to $K(\zeta, h)$ and satisfies (2.17). This completes the proof in this case. Note that this is the case when the disk does not touch the wire. Suppose now that $h > -r$. Then, the function u defined by (2.24) belongs to K. Moreover, it holds that:

$$
\int_\Omega u_x(v_x - u_x) \, dx = \int_0^{x_1} u_x(v_x - u_x) \, dx + \int_{x_1}^{y_1} u_x(v_x - u_x) \, dx
$$
$$
+ \int_{y_1}^1 u_x(v_x - u_x) \, dx. \tag{2.25}
$$

Writing down in each intervals

$$u_x(v_x - u_x) = (u_x(v - u))_x - u_{xx}(v - u),$$

it comes after integration

$$
\int_\Omega u_x(v_x - u_x) \, dx = u_x(v - u) \big|_0^{x_1} + u_x(v - u) \big|_{x_1}^{y_1}
$$
$$
- \int_{x_1}^{y_1} \psi_{xx}(v - \psi) \, dx + u_x(v - u) \big|_{y_1}^1.
$$

(We used the fact that $u_{xx} = 0$ on $(0, x_1)$, $(y_1, 1)$). Since u, v vanish on the boundary of Ω and since u is clearly a C^1-function it comes

$$\int_\Omega u_x (v_x - u_x) \, dx = -\int_{x_1}^{y_1} \psi_{xx}(v - \psi) \, dx \quad \forall v \in K(\zeta, h). \tag{2.26}$$

Since $\psi_{xx} \leq 0$, $v \geq \psi$ on (x_1, y_1) it comes

$$u \in K(\zeta, h), \qquad \int_\Omega u_x (v_x - u_x) \, dx \geq 0 \quad \forall v \in K(\zeta, h), \tag{2.27}$$

and u defined by (2.24) is solution to (2.16), (2.17). This completes the proof of the proposition.   □

Remark 2.2. The variational inequality (2.17) is called an obstacle problem (see [5], [15], [19]). The set

$$[x_1, y_1] = \{ x \in \Omega \mid u(x) = \psi^{\zeta h}(x) \} \tag{2.28}$$

is called the "coincidence set" of this obstacle problem.

One should also notice that it holds that

$$\begin{cases} u_{xx} \leq 0 & \text{in } \Omega, \\ u_{xx} = 0 & \text{in } \Omega \setminus [x_1, y_1], \\ u = \psi & \text{in } [x_1, y_1]. \end{cases} \tag{2.29}$$

(h is directed downward.)

Let us now come back to our problem (2.9). Due to (2.14) and Proposition 2.1 we see that if a solution exists then

$$u_0 = u_0(\zeta_0, h_0)$$

and it holds that

$$E(u(\zeta, h); \zeta, h) \leq E(u; \zeta, h) \quad \forall u \in K(\zeta, h). \tag{2.30}$$

Thus, for $u = u(\zeta, h)$ let us set

$$e(\zeta, h) = \frac{1}{2} \int_\Omega u_x^2 \, dx - Gh. \tag{2.31}$$

Due to (2.30), the problem (2.9) reduces to minimize $e(\zeta, h)$ on $(r, 1 - r) \times \mathbb{R}$, i.e. to a minimization problem in two dimensions. However, the set where we are minimizing is not compact and we do not know yet if the function $e = e(\zeta, h)$ is continuous. This is what we would like to address now. For that let us find the expressions of x_1, y_1 in term of ζ, h. We have

Lemma 2.1. *Let us assume $h > -r$ and the notation of Figure 6. Then it holds that*

$$x_1 = x_1(\zeta, h) = \zeta - \frac{r^2\zeta + rh\sqrt{h^2 + \zeta^2 - r^2}}{h^2 + \zeta^2}, \tag{2.32}$$

$$y_1 = y_1(\zeta, h) = \zeta + \frac{r^2(1 - \zeta) + rh\sqrt{h^2 + (1 - \zeta)^2 - r^2}}{h^2 + (1 - \zeta)^2}. \tag{2.33}$$

In particular, x_1, y_1 are differentiable.

Proof. Since $0X_1$ and X_1C are orthogonal it holds that

$$x_1(x_1 - \zeta) + \psi(x_1)(\psi(x_1) - h) = 0,$$

$$\Longleftrightarrow \quad x_1(x_1 - \zeta) + \{h + \sqrt{r^2 - (x_1 - \zeta)^2}\}\sqrt{r^2 - (x_1 - \zeta)^2} = 0,$$

$$\Longleftrightarrow \quad \zeta(x_1 - \zeta) + r^2 + h\sqrt{r^2 - (x_1 - \zeta)^2} = 0.$$

Setting $X = \zeta - x_1$ we get

$$-\zeta X + r^2 + h\sqrt{r^2 - X^2} = 0$$

$$\Longleftrightarrow \quad h^2(r^2 - X^2) = (r^2 - \zeta X)^2, \quad (\zeta X - r^2) \cdot h \geq 0. \tag{2.34}$$

Expanding the square in the first equation it comes:

$$X^2(\zeta^2 + h^2) - 2r^2\zeta X + r^4 - r^2h^2 = 0 \tag{2.35}$$

$$\Longleftrightarrow \quad X^2 - \frac{2r^2\zeta X}{\zeta^2 + h^2} + \frac{r^4 - r^2h^2}{\zeta^2 + h^2} = 0$$

$$\Longleftrightarrow \quad \left(X - \frac{r^2\zeta}{\zeta^2 + h^2}\right)^2 + \frac{(r^4 - r^2h^2)(\zeta^2 + h^2) - r^4\zeta^2}{(\zeta^2 + h^2)^2} = 0$$

$$\Longleftrightarrow \quad \left(X - \frac{r^2\zeta}{\zeta^2 + h^2}\right)^2 = \frac{r^2h^2\{\zeta^2 + h^2 - r^2\}}{(\zeta^2 + h^2)^2}$$

$$\Longleftrightarrow \quad X = \frac{r^2\zeta \pm rh\sqrt{\zeta^2 + h^2 - r^2}}{\zeta^2 + h^2}. \tag{2.36}$$

(Recall that $\zeta \in (r, 1 - r)$ so that the term under the square root symbol is positive.) Going back to (2.34) we see that

$$\zeta X = r^2 \pm h\sqrt{r^2 - X^2}, \quad (\zeta X - r^2) \cdot h \geq 0$$

so that X is the largest root of the equation (2.35) for $h > 0$ and the smallest for $h < 0$. Thus $X = \zeta - x_1$ is given by (2.36) with the sign $+$. This leads to (2.32). The proof of (2.33) is similar changing ζ into $1 - \zeta$ and this is left to the reader. $\qquad \square$

In order to minimize the function $e(\zeta, h)$ we first study its differentiability. We have then:

Theorem 2.1. *The function* $e = e(\zeta, h)$ *defined by Equation* (2.31) *is differentiable on* $(r, 1 - r) \times \mathbb{R}$ *and it holds that:*

- *For* $h \leq -r$

$$\frac{\partial e}{\partial \zeta}(\zeta, h) = 0, \quad \frac{\partial e}{\partial h}(\zeta, h) = -G, \tag{2.37}$$

- *For* $h > -r$

$$\frac{\partial e}{\partial \zeta}(\zeta, h) = \frac{1}{2}\{\psi_x^2(y_1) - \psi_x^2(x_1)\},$$

$$\frac{\partial e}{\partial h}(\zeta, h) = \psi_x(x_1) - \psi_x(y_1) - G. \tag{2.38}$$

Here x_1, y_1 *are the points given by* (2.32), (2.33) *and* ψ_x *denotes the derivative in x of the function* ψ.

Proof. a) *A formal proof.*

The rigorous proof of (2.38) being tedious let us give first a formal proof of it. We restrict ourselves to the first formula of (2.38). Due to (2.31) – if every differentiation under the integral sign is justified – we have

$$\frac{\partial e}{\partial \zeta}(\zeta, h) = \frac{1}{2}\frac{\partial}{\partial \zeta}\int_\Omega u_x^2 \, dx = \int_\Omega u_x u_{x\zeta} \, dx \tag{2.39}$$

with an obvious notation for $u_{x\zeta}$. Now, clearly, we have

$$u = u(\zeta, h) = 0 \quad \text{at } 0, 1 \quad \forall \zeta, h. \tag{2.40}$$

(Recall that $u = u(\zeta, h)$ is the solution of the variational inequality (2.17)). Thus it holds that

$$u_\zeta = 0 \quad \text{at } 0, 1 \quad \forall \zeta, h. \tag{2.41}$$

Integrating the last integral of (2.39) by parts we get

$$\frac{\partial e}{\partial \zeta}(\zeta, h) = -\int_0^1 u_{xx} u_\zeta \, dx.$$

Due to (2.29) we obtain then

$$\frac{\partial e}{\partial \zeta}(\zeta, h) = -\int_{x_1}^{y_1} \psi_{xx} \psi_\zeta \, dx. \tag{2.42}$$

According to (2.3) we have

$$\psi_\zeta = -\psi_x$$

so that (2.42) becomes

$$\frac{\partial e}{\partial \zeta}(\zeta, h) = \int_{x_1}^{y_1} \psi_{xx} \psi_x \, dx = \frac{1}{2} \{\psi_x(y_1)^2 - \psi_x(x_1)^2\}$$

which is the first formula (2.38).

b) *The rigorous proof.*

First, due to (2.23), (2.31) for $h \le -r$ we have

$$e(\zeta, h) = -Gh \qquad (2.43)$$

and thus (2.37) holds in this case.

Let us assume that $h > -r$. Due to (2.24), (2.31) we have

$$e(\zeta, h) = \frac{1}{2} \int_0^{x_1} u_x^2 \, dx + \frac{1}{2} \int_{x_1}^{y_1} u_x^2 \, dx + \frac{1}{2} \int_{y_1}^1 u_x^2 \, dx - Gh$$

$$= \frac{1}{2} x_1 \psi_x^2(x_1) + \frac{1}{2} \int_{x_1}^{y_1} \psi_x^2 \, dx + \frac{1}{2} (1 - y_1) \psi_x^2(y_1) - Gh. \qquad (2.44)$$

Let us set

$$A = \frac{1}{2} \int_{x_1}^{y_1} \psi_x^2 \, dx. \qquad (2.45)$$

Applying the chain rule we get

$$\frac{\partial A}{\partial \zeta}(\zeta, h) = -\frac{1}{2} \psi_x^2(x_1) \frac{\partial x_1}{\partial \zeta} + \int_{x_1}^{y_1} \psi_x \psi_{x\zeta} \, dx + \frac{1}{2} \psi_x(y_1)^2 \frac{\partial y_1}{\partial \zeta}. \qquad (2.46)$$

(Recall that by (2.32), (2.33) x_1 and y_1 are differentiable and so is ψ).

Since by (2.3)

$$\psi_{x\zeta} = -\psi_{xx},$$

(2.46) can be written

$$\frac{\partial A}{\partial \zeta} = -\frac{1}{2} \psi_x^2(x_1) \frac{\partial x_1}{\partial \zeta} + \frac{1}{2} \psi_x(x_1)^2 - \frac{1}{2} \psi_x(y_1)^2 + \frac{1}{2} \psi_x(y_1)^2 \frac{\partial y_1}{\partial \zeta}. \qquad (2.47)$$

Going back to (2.44) and differentiating e we obtain

$$\frac{\partial e}{\partial \zeta}(\zeta, h) = \frac{1}{2} \psi_x^2(x_1) \frac{\partial x_1}{\partial \zeta} + x_1 \psi_x(x_1) \frac{\partial \psi_x(x_1)}{\partial \zeta} + \frac{\partial A}{\partial \zeta}$$

$$\qquad - \frac{1}{2} \psi_x(y_1)^2 \frac{\partial y_1}{\partial \zeta} + (1 - y_1) \psi_x(y_1) \frac{\partial \psi_x(y_1)}{\partial \zeta}$$

$$= x_1 \psi_x(x_1) \frac{\partial \psi_x(x_1)}{\partial \zeta} + \frac{1}{2} \psi_x(x_1)^2$$

$$\qquad - \frac{1}{2} \psi_x(y_1)^2 + (1 - y_1) \psi_x(y_1) \frac{\partial \psi_x(y_1)}{\partial \zeta}, \qquad (2.48)$$

by (2.47). We remark then that

$$x_1 \psi_x(x_1) = \psi(x_1) \quad \text{and} \quad (y_1 - 1)\psi_x(y_1) = \psi(y_1), \tag{2.49}$$

hold for any (ζ, h). Thus differentiating in ζ these equalities we obtain

$$\psi_x(x_1)\frac{\partial x_1}{\partial \zeta} + x_1 \frac{\partial \psi_x(x_1)}{\partial \zeta} = \psi_\zeta(x_1) + \psi_x(x_1)\frac{\partial x_1}{\partial \zeta}, \tag{2.50}$$

$$\psi_x(y_1)\frac{\partial y_1}{\partial \zeta} + (y_1 - 1)\frac{\partial \psi_x(y_1)}{\partial \zeta} = \psi_\zeta(y_1) + \psi_x(y_1)\frac{\partial y_1}{\partial \zeta}. \tag{2.51}$$

Thus it holds

$$x_1 \frac{\partial \psi_x(x_1)}{\partial \zeta} = \psi_\zeta(x_1) = -\psi_x(x_1), \quad (y_1 - 1)\frac{\partial \psi_x(y_1)}{\partial \zeta} = \psi_\zeta(y_1) = -\psi_x(y_1).$$

Reporting this to (2.48) we obtain the first formula of (2.38). To get the second formula we proceed similarly. We have first by the chain rule and (2.45)

$$\frac{\partial A}{\partial h} = \frac{\partial A}{\partial h}(\zeta, h) = -\frac{1}{2}\psi_x(x_1)^2\frac{\partial x_1}{\partial h} + \int_{x_1}^{y_1} \psi_x \psi_{xh}\, dx + \frac{1}{2}\psi_x(y_1)^2\frac{\partial y_1}{\partial h}. \tag{2.52}$$

Since $\psi_{xh} = \psi_{hx} = 0$ it comes

$$\frac{\partial A}{\partial h} = -\frac{1}{2}\psi_x(x_1)^2\frac{\partial x_1}{\partial h} + \frac{1}{2}\psi_x(y_1)^2\frac{\partial y_1}{\partial h}. \tag{2.53}$$

Differentiating e given by (2.44) in the h direction we obtain

$$\begin{aligned}
\frac{\partial e}{\partial h}(\zeta, h) &= \frac{1}{2}\psi_x^2(x_1)\frac{\partial x_1}{\partial h} + x_1 \psi_x(x_1)\frac{\partial \psi_x(x_1)}{\partial h} + \frac{\partial A}{\partial h} \\
&\quad - \frac{1}{2}\psi_x^2(y_1)\frac{\partial y_1}{\partial h} + (1 - y_1)\psi_x(y_1)\frac{\partial \psi_x(y_1)}{\partial h} - G \\
&= x_1 \psi_x(x_1)\frac{\partial \psi_x(x_1)}{\partial h} + (1 - y_1)\psi_x(y_1)\frac{\partial \psi_x(y_1)}{\partial h} - G
\end{aligned} \tag{2.54}$$

by (2.53). Differentiating then (2.49) with respect to h we obtain

$$\psi_x(x_1)\frac{\partial x_1}{\partial h} + x_1 \frac{\partial \psi_x(x_1)}{\partial h} = \psi_h(x_1) + \psi_x(x_1)\frac{\partial x_1}{\partial h},$$

$$\psi_x(y_1)\frac{\partial y_1}{\partial h} + (y_1 - 1)\frac{\partial \psi_x(y_1)}{\partial h} = \psi_h(y_1) + \psi_x(y_1)\frac{\partial y_1}{\partial h}$$

and thus

$$x_1 \frac{\partial \psi_x(x_1)}{\partial h} = \psi_h(x_1) = 1, \quad (y_1 - 1)\frac{\partial \psi_x}{\partial h}(y_1) = \psi_h(y_1) = 1. \tag{2.55}$$

Using that in (2.54) leads to the second formula of (2.38). Now, clearly when $h > -r$, $h \to -r$, $\psi_x(x_1), \psi_x(y_1) \to 0$ and thus these partial derivatives of e are continuous in $(r, 1 - r) \times \mathbb{R}$. The differentiability of e follows and the proof of the theorem is complete. □

We can now solve the problem (2.9). We have

Theorem 2.2. *The problem (2.9) admits a unique solution given by*

$$\zeta_0 = \frac{1}{2}, \quad h_0 = \frac{G}{4} - r\sqrt{1 + \frac{G^2}{4}}, \quad u_0 = u(\zeta_0, h_0) \qquad (2.56)$$

where $u(\zeta_0, h_0)$ *is the solution of the variational inequality (2.17) given by (2.24).*

Proof. First let us remark that, as we could expect, the disk reaches its equilibrium when centered in the middle of the wire.

As observed below (2.31) the problem (2.9) reduces to minimize the function $e(\zeta, h)$ on $(r, 1 - r) \times \mathbb{R}$. Let us consider first $h > -r$. Then, due to (2.38) we have

$$\frac{\partial e}{\partial \zeta}(\zeta, h) < 0 \quad \forall \zeta < \frac{1}{2}, \qquad \frac{\partial e}{\partial \zeta}(\zeta, h) > 0 \quad \forall \zeta > \frac{1}{2}. \qquad (2.57)$$

Thus in this case the minimum of $e(\zeta, h)$ is clearly achieved for $\zeta = \frac{1}{2}$. For $h \le -r$ we have

$$e(\zeta, h) = -Gh \qquad (2.58)$$

i.e. e is independent of ζ and of course the minimum of e is achieved also for $\zeta = \zeta_0 = \frac{1}{2}$. Thus, we have now reduced the problem to minimize $e(\frac{1}{2}, h)$ over $\mathbb{R}$. For $h > -r$ one has

$$\frac{d}{dh}\left\{e\left(\frac{1}{2}, h\right)\right\} = \frac{\partial e}{\partial h}\left(\frac{1}{2}, h\right) = \psi_x(x_1) - \psi_x(y_1) - G = 2\psi_x(x_1) - G \qquad (2.59)$$

since for symmetry reasons $\psi_x(y_1) = -\psi_x(x_1)$.

Thus, we have

$$\frac{d}{dh}\left\{e\left(\frac{1}{2}, h\right)\right\} < 0 \quad \text{for } \psi_x(x_1) < \frac{G}{2},$$

$$\frac{d}{dh}\left\{e\left(\frac{1}{2}, h\right)\right\} > 0 \quad \text{for } \psi_x(x_1) > \frac{G}{2} \qquad (2.60)$$

and a unique minimum is achieved on $(-r, +\infty)$ when it holds that

$$\psi_x(x_1) = -\psi_x(y_1) = \frac{G}{2}. \qquad (2.61)$$

Since e is continuous for $h = -r$, this minimum is also the minimum of $e(\frac{1}{2}, h)$ on the whole real line. This completes the proof of existence and uniqueness of a unique minimizer (u_0, ζ_0, h_0). To determine h_0, if we set

$$\frac{G}{2} = \tan g\,\theta \qquad (2.62)$$

then we have (see Figure 7)

$$h_0 = x_1 \frac{G}{2} - r \sin\left(\frac{\pi}{2} - \theta\right) = x_1 \frac{G}{2} - r \cos\theta, \tag{2.63}$$

$$x_1 = \zeta - r \cos\left(\frac{\pi}{2} - \theta\right) = \frac{1}{2} - r \sin\theta. \tag{2.64}$$

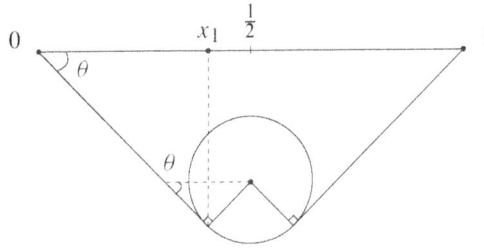

Figure 7. The equilibrium position of a single disk

From this it follows that

$$h_0 = \frac{G}{4} - \frac{G}{2} r \sin\theta - r \cos\theta \tag{2.65}$$

with (by (2.62))

$$\sin\theta = \frac{G}{2} \cos\theta.$$

Since

$$1 = \sin^2\theta + \cos^2\theta = \left(1 + \frac{G^2}{4}\right)\cos^2\theta,$$

we obtain

$$\cos\theta = \frac{1}{\sqrt{1 + \frac{G^2}{4}}}, \quad \sin\theta = \frac{G}{2} \cdot \frac{1}{\sqrt{1 + \frac{G^2}{4}}}$$

and by (2.65)

$$h_0 = \frac{G}{4} - r\sqrt{1 + \frac{G^2}{4}}. \tag{2.66}$$

This completes the proof of the theorem.   □

2.2 The case of one square on a wire

In this section we would like to address the case of a square moving freely on an elastic wire. We will not provide all the details of the analysis in this case referring the interested reader to [2], but we would like to show that a surprising behaviour can happen in this case.

The setting is the same as in the previous section. We suppose that in its undeformed configuration the elastic wire occupies the domain $\Omega = (0, 1)$. We denote by $2r$ the length of the sides of the square and by G its weight.

If Q denotes the closed set of the points occupied by the square we will always assume that

$$Q \subset \Omega \times \mathbb{R}. \tag{2.67}$$

We will denote by $P = (\zeta_p, h_p)$ the barycenter of the square – see Figure 8. For

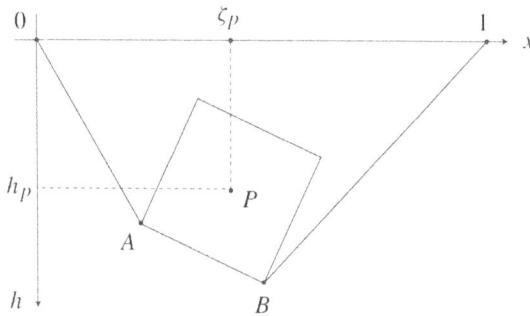

Figure 8. A square on a wire

$u \in H_0^1(\Omega)$ we set

$$C_u = \{ (x, h) \in \Omega \times \mathbb{R} \mid h \leq u(x) \}. \tag{2.68}$$

(Recall that a function in $H_0^1(\Omega)$ is a continuous function in this case – see [6]). Then, for a position Q of the square, an admissible deformation of the wire is a function u in $H_0^1(\Omega)$ such that

$$Q \subset C_u. \tag{2.69}$$

The energy of an admissible deformation is given by the same formula as in the preceding section. Then we would like to find

$$\operatorname*{Inf}_{(Q,u)} \frac{1}{2} \int_\Omega u_x^2(x)\,dx - Gh_p^Q \tag{2.70}$$

when Q is located in the strip $\Omega \times \mathbb{R}$ and when $u \in H_0^1(\Omega)$ is such that (2.69) holds. h_p^Q denotes the second coordinate of the barycenter P of the square occupying the position Q. More precisely we would like to show that the problem (2.70) admits a solution which is unique up to symmetry. As for the case of a disk we need the square to be able to fit inside the interval (0, 1) and in order to insure that we suppose

$$r < \frac{1}{4}. \tag{2.71}$$

In (2.70) we are minimizing over a set of isometric squares and admissible deformations so that the situation seems slightly unusual. As we have done in the case of a disk we are going to reformulate the problem as a minimization problem on a subset of $\mathbb{R}^4$.

Suppose that the position of the square Q is fixed in the strip $\Omega \times \mathbb{R}$. Denote by $I_Q = \Pi_x(Q)$ the interval of Ω projection of Q on the x-axis (Π_x is the orthogonal projection on the x-axis). For $x \in I_Q$ set

$$\psi_Q(x) = \text{Sup}\{ y \mid (x, y) \in Q\}. \tag{2.72}$$

Then, clearly, the function ψ_Q is a function describing the lower border of Q. Then, with this notation, the constraint (2.69) can be written

$$u \in C_Q = \{ u \in H_0^1(\Omega) \mid u(x) \geq \psi_Q(x) \text{ on } I_Q \}. \tag{2.73}$$

It is clear if (Q_0, u_0) is a minimizer (2.70) that it holds that

$$u_0 \in C_{Q_0}, \quad \frac{1}{2} \int_\Omega u_{0x}^2 \, dx \leq \frac{1}{2} \int_\Omega u_x^2 \, dx \quad \forall u \in C_{Q_0}, \tag{2.74}$$

i.e. u_0 is solution of the variational inequality

$$\begin{cases} u_0 \in C_{Q_0}, \\ \int_\Omega u_{0x}(v_x - u_{0x}) \, dx \geq 0 \quad \forall v \in C_{Q_0}. \end{cases} \tag{2.75}$$

(It is easy to see that C_{Q_0} is a closed convex set of $H_0^1(\Omega)$. As in the case of a disk, u_0 is the projection of 0 on C_{Q_0} – see Proposition 2.1.) Thus, for Q satisfying (2.67) we denote by $u = u_Q$ the unique solution to

$$\begin{cases} u \in C_Q, \\ \int_\Omega u_x(v_x - u_x) \, dx \geq 0 \quad \forall v \in C_Q. \end{cases} \tag{2.76}$$

It is then clear that (2.70) is equivalent to

$$\underset{Q \subset \Omega \times \mathbb{R}}{\text{Inf}} \frac{1}{2} \int_\Omega u_{Qx}^2 \, dx - Gh_p^Q, \tag{2.77}$$

i.e., we minimize on Q only. Now, a square is perfectly determined by the position of two of its vertices – for instance by A, B as in Figure 8. A being the further left vertex and B the further down. If we set

$$A = (x_A, y_A), \quad B = (x_B, y_B) \tag{2.78}$$

we can always assume that

$$0 < x_A < x_B < 1, \quad y_A \leq y_B \tag{2.79}$$

(recall that the vertical axis is directed downward). Now, the two points A, B are not able to move freely. They always have to fulfill the constraint

$$(x_A - x_B)^2 + (y_A - y_B)^2 = 4r^2. \tag{2.80}$$

If we choose A, B satisfying (2.79), (2.80) the square Q is perfectly determined and we have

$$\frac{1}{2} \int_\Omega u_{Qx}^2 \, dx - Gh_p^Q = e(x_A, y_A, x_B, y_B) \tag{2.81}$$

i.e. the energy that we have to minimize is a function of the points belonging to the subset of $\mathbb{R}^4$ defined by (2.79), (2.80). Then, using a compactness argument we can show

Theorem 2.3. *The problems (2.70), (2.77) admit a minimizer.*

Proof. As seen above, it is enough to minimize (2.81). We refer the reader to [2] for the details of this minimization. □

Now when it comes to find minimizers for (2.70), (2.77) the first guess is to think that the square will find its equilibrium by sitting horizontally in the middle of the wire. It is true for small values of the weight of the square. Surprisingly this holds no more when G becomes large. In other words, the horizontal position of the square becomes unstable when the tension of the wire increases. We have indeed

Theorem 2.4. *Assume that*

$$G \leq 2(\sqrt{3} - 1) \tag{2.82}$$

then the problems (2.70), (2.77) admit a unique solution given by a square centered in the middle of the wire (see Figure 9) with

$$x_A = \frac{1}{2} - r, \quad x_B = \frac{1}{2} + r, \quad y_A = y_B = \frac{G}{2}\left(\frac{1}{2} - r\right). \tag{2.83}$$

Assume that

$$2(\sqrt{3} - 1) < G < 2 \tag{2.84}$$

Michel Chipot

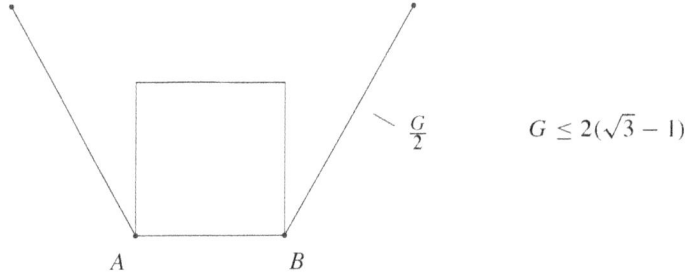

$$G \leq 2(\sqrt{3} - 1)$$

Figure 9. The square sitting in the middle

then the problems (2.70), (2.77) *admit two solutions. One is given by A, B satisfying*

$$x_A = \frac{1}{2} - \frac{r}{\sqrt{G^2 + 4G - 4}}\left(2 + \frac{2\sqrt{G^2 + 4G - 8}}{G}\right), \tag{2.85}$$

$$y_A = \frac{G}{4} - \frac{r}{\sqrt{G^2 + 4G - 4}}\left(G + \sqrt{G^2 + 4G - 8}\right), \tag{2.86}$$

$$x_B = \frac{1}{2} + \frac{r}{\sqrt{G^2 + 4G - 4}}\left(2 - \frac{2\sqrt{G^2 + 4G - 8}}{G}\right), \tag{2.87}$$

$$y_B = \frac{G}{4} - \frac{r}{\sqrt{G^2 + 4G - 4}}\left(2 - \sqrt{G^2 + 4G - 8}\right). \tag{2.88}$$

The other solution is obtained by reflection with respect to the vertical axis $x = \frac{1}{2}$
(*see Figure* 10).

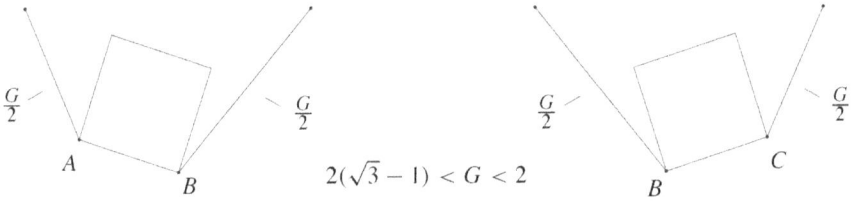

$$2(\sqrt{3} - 1) < G < 2$$

Figure 10. The two disks are slightly tilted

Assume that

$$2 \leq G \leq 2\sqrt{7} \tag{2.89}$$

then the problems (2.70), (2.77) admit a unique minimizer given by a square centered in the middle of the wire with A, B given by

$$x_A = \frac{1}{2} - \sqrt{2}r, \quad y_A = \frac{G}{2}\left(\frac{1}{2} - \sqrt{2}r\right), \quad x_B = \frac{1}{2}, \quad y_B = \frac{G}{4} + \sqrt{2}r\left(1 - \frac{G}{2}\right) \quad (2.90)$$

(see Figure 11).

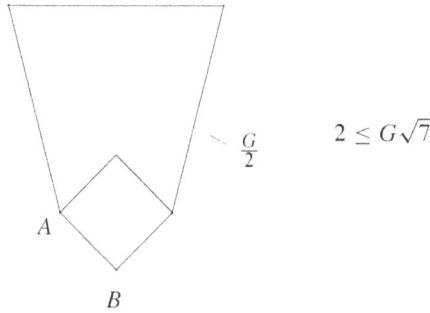

Figure 11. The square is centered again

Assume now that

$$2\sqrt{7} < G \quad (2.91)$$

then the problems (2.70) or (2.77) admit two symmetric solutions – see Figure 12.

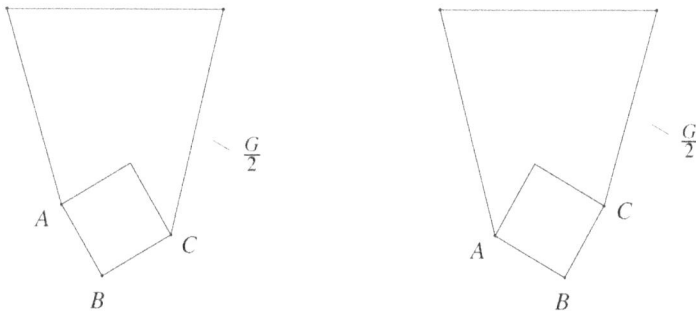

Figure 12. The square tilts again for large G

Proof. We refer to [2]. Let us just mention that the proof relies on Lagrange multipliers technique. In the case of (2.91) it is also possible to determine A and B exactly. However, the expressions are complicated. They are given in [1], [2]. □

Michel Chipot

2.3 Some open questions

As we mentioned in Remark 2.1 one can consider the case of an inhomogeneous
wire and the case of an inhomogeneous two dimensional body sitting above it: for
instance a disk, a square... In the homogeneous case – for an object or a wire – one
can consider the case of a rectangle which should be a variant of the case of a square.
More challenging is the case of an ellipse. Indeed in this case, depending of the weight
of it the equilibrium positions should be like in Figure 13 for light weights and like in
Figure 14 for large weights, i.e., when the ellipse is light then it should be lying with
its long axis horizontal and on the contrary with its long axis vertical when the weight
is large. (Note that the scales of the figures are not the same.) Of course the example
of the square tells us that in between or for very large weights a tilted position could
also occur. We leave these issues to the reader's curiosity.

Figure 13. The case of an
ellipse with light weight

Figure 14. The case of an
ellipse with large weight

Another variant of the above problem is to consider an energy related to the arc
length of the deformation (which would correspond to the mean curvature operator in
higher dimensions). In this case one would minimize

$$E = \frac{1}{2} \int_{\Omega} \sqrt{1 + a(x)u_x^2(x)} \, dx - Gh \tag{2.92}$$

and all our analysis has to be revisited.

Of course the notation of equilibrium refers also to an evolution process, i.e. the
equilibrium reached for instance by the disk corresponds to the final state of its motion
when we drop it on the wire and let him roll freely. This has to be established mathematically and in particular the evolution problem has to be defined and considered
together with its asymptotic behaviour.

3 The case of two disks on an elastic wire

3.1 Formulation of the problem

We consider two solid disks of different sizes and weights allowed to roll freely on an elastic wire – see Figure 15.

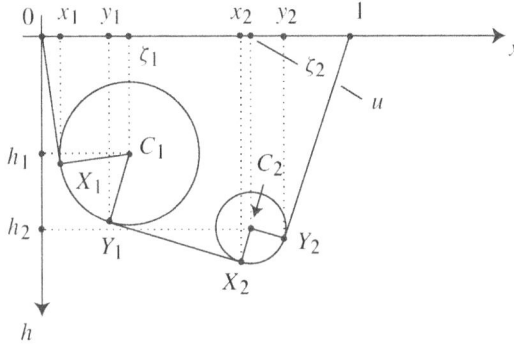

Figure 15. Two disks rolling on a wire

We denote by r_1, r_2 the different radii of these disks and by G_1, G_2 their respective weights. In its undeformed state the elastic wire is supposed to occupy the interval $\Omega = (0, 1)$. Letting roll the two disks on it, they will reach an equilibrium position that we would like to determine. First, as we did before for one single disk we notice that the position of the disks is completely determined by the location of their centers

$$C_1 = (\zeta_1, h_1), \quad C_2 = (\zeta_2, h_2). \tag{3.1}$$

If u is an admissible deformation of the wire – we will come back to this point later – the energy of the configuration is given by

$$E = \frac{1}{2} \int_\Omega u_x^2 \, dx - G_1 h_1 - G_2 h_2. \tag{3.2}$$

Our goal in studying this one dimensional problem is to get some insight on what happens in higher dimension that is to say for instance for two balls rolling on a membrane.

First, we will assume that both disks are located in the strip $\Omega \times \mathbb{R}$, that is to say, that

$$\zeta_i \in (r_i, 1 - r_i), \quad i = 1, 2. \tag{3.3}$$

Next, we will suppose that

$$(\zeta_1 - \zeta_2)^2 + (h_1 - h_2)^2 \geq (r_1 + r_2)^2 \tag{3.4}$$

i.e. we will not allow the disks to interpenetrate each other.

The lower part of each disk can be described by the function

$$\psi_i = \psi^{\zeta_i h_i}(x) = h_i + \sqrt{r_i^2 - (x - \zeta_i)^2}, \quad i = 1, 2, \tag{3.5}$$

defined on $B(\zeta_i, r_i) = (\zeta_i - r_i, \zeta_i + r_i)$. Then, an admissible deformation is a function u such that the disks are located below u – note that we suppose the vertical direction h oriented downward (see Figure 15). More precisely an admissible deformation is a function u such that

$$u \in K = K(C_1, C_2) = \{ v \in H_0^1(\Omega) \mid v(x) \geq \psi_i(x) \text{ on } B(\zeta_i, r_i) \, i = 1, 2 \}. \tag{3.6}$$

Then the problem we have to consider is to find

$$\underset{(u, C_1, C_2)}{\text{Inf}} \frac{1}{2} \int_\Omega u_x^2 \, dx - G_1 h_1 - G_2 h_2 \tag{3.7}$$

where u belongs to K and C_1, C_2 the center of the disks are satisfying the constraints (3.3), (3.4).

We would like to show first that one can reformulate the problem (3.7) as a minimization problem in $\mathbb{R}^4$. Assume that we have found an element $(u^0, C_1^0, C_2^0) = (u^0, \zeta_1^0, h_1^0, \zeta_2^0, h_2^0)$ minimizing (3.7). Then, clearly, u^0 minimizes

$$\frac{1}{2} \int_\Omega v_x^2 \, dx - G_1 h_1^0 - G_2 h_2^0$$

on the set

$$K^0 = K^0(\zeta_1^0, h_1^0, \zeta_2^0, h_2^0) = \{ v \in H_0^1(\Omega) \mid v(x) \geq \psi_i(x) \text{ on } B(\zeta_i^0, r_i), \, i = 1, 2 \}$$

i.e. since $-G_1 h_1^0 - G_2 h_2^0$ is independent of v, u^0 is the solution of the variational inequality

$$\begin{cases} u^0 \in K^0, \\ \int_\Omega u_x^0 (v_x - u_x^0) \, dx \geq 0 \quad \forall v \in K^0. \end{cases}$$

Thus, for $\xi = (\zeta_1, h_1, \zeta_2, h_2) \in \mathbb{R}^4$ we introduce $u = u(\xi)$ the solution of the variational inequality

$$\begin{cases} u \in K(\xi) = \{ v \in H_0^1(\Omega) \mid v(x) \geq \psi_i(x) \text{ on } B(\zeta_i, r_i), \, i = 1, 2 \}, \\ \int_\Omega u_x (v_x - u_x) \, dx \geq 0 \quad \forall v \in K(\xi), \end{cases} \tag{3.8}$$

and we set for $u = u(\xi)$

$$e(\xi) = e(\zeta_1, h_1, \zeta_2, h_2) = \frac{1}{2} \int_\Omega u_x^2 \, dx - G_1 h_1 - G_2 h_2. \tag{3.9}$$

The problem (3.7) is then clearly equivalent to:

$$\begin{cases} \text{find } \xi^0 = (\zeta_1^0, h_1^0, \zeta_2^0, h_2^0) \text{ satisfying } (3.3), (3.4) \text{ such that} \\ e(\xi^0) \leq e(\xi) \quad \forall \xi = (\zeta_1, h_1, \zeta_2, h_2) \text{ satisfying } (3.3), (3.4). \end{cases} \tag{3.10}$$

Thus, our problem is reformulated as a problem of minimization on $\mathbb{R}^4$. Note that the set of points where we are minimizing – see (3.3) – is not compact and even if e is continuous, as before, some work has to be done to obtain the existence of a minimizer.

First – and without loss of generality – we will assume all along

$$r_1 \geq r_2. \tag{3.11}$$

Moreover, we will suppose that when one of the disks is sitting on the wire we can always fit the other one on one side or the other. In other words we will suppose

$$2r_1 + r_2 < \frac{1}{2}. \tag{3.12}$$

3.2 Preparatory results

Let us denote by $D_i = D_i(C_i, r_i), i = 1, 2$ the closed disk of center $C_i = (\zeta_i, h_i)$ and radius r_i. Due to our assumption (3.3) the two disks are located in the strip $(0, 1) \times \mathbb{R}$. Denote then by U the convex hull in $\mathbb{R}^2$ of $D_1 \cup D_2 \cup [0, 1] \times (-\infty, 0]$. Then for $x \in [0, 1]$ set

$$u(x) = \text{Sup}\{ y \mid (x, y) \in U \}. \tag{3.13}$$

One has:

Proposition 3.1. *The function u defined by (3.13) coincides with the solution $u = u(\xi) = u(\zeta_1, h_1, \zeta_2, h_2)$ of the variational inequality (3.8).*

Proof. In the case $D_1, D_2 \subset (0, 1) \times (-\infty, 0]$ one has

$$u = 0$$

and $u = 0$ satisfies clearly (3.8) in this case.

Let us assume now that we are in the case of Figure 15 – which is somehow the generic case. $u = u(\xi)$ is the function shown on the picture. Then it holds – with the notation of Figure 15 – for any $v \in K(\xi)$:

$$\int_0^1 u_x(v_x - u_x) \, dx = \int_0^1 (u_x(v - u))_x - u_{xx}(v - u) \, dx, \tag{3.14}$$

where the second integral is understood as the sum of the integrals on $(0, x_1), (x_1, y_1), (y_1, x_2), (x_2, y_2), (y_2, 1)$ intervals on which u is C^2. The first integral of the second side of (3.14) vanishes. Indeed this follows after summing the equalities:

$$\int_0^{x_1} (u_x(v - u))_x \, dx = u_x(v - u) \Big|_0^{x_1} = u_x(v - u)(x_1),$$

$$\int_{x_1}^{y_1} (u_x(v - u))_x \, dx = u_x(v - u) \Big|_{x_1}^{y_1} = u_x(v - u)(y_1) - u_x(v - u)(x_1), \ldots$$

. . . . . .

(We use the fact that u is C^1, $v - u = 0$ on $\{0, 1\}$.) Thus, from (3.14) it follows that

$$\int_0^1 u_x(v - u_x) \, dx = -\int_{x_1}^{y_1} \psi_{1.xx}(v - \psi_1) \, dx - \int_{x_2}^{y_2} \psi_{2.xx}(v - \psi_2) \, dx \tag{3.15}$$

$$\geq 0 \qquad \forall v \in K(\xi).$$

(It is due to the fact that $u_{xx} = 0$ outside (x_i, y_i) and the fact that $\psi_{ixx} \leq 0$, $v \geq \psi_i$ on (x_i, y_i), $i = 1, 2$).

This completes the proof in this case. In any other case one proceeds with the same arguments splitting the integral into the different intervals where u is of class C^2. Note that this works also when the disks are interpenetrating each other. The details are left to the reader.   □

Remark 3.1. What we call above the other cases is for instance when only one disk is touching the membrane, or when the small one is located below the large one – see below Figure 17.

Remark 3.2. We can define the coincidence sets

$$I_i = \{ x \in (0, 1) \mid u(x) = \psi_i(x) \}, \quad i = 1, 2. \tag{3.16}$$

Then, the solution u satisfies clearly

$$-u_{xx} \geq 0 \quad \text{in } (0, 1), \tag{3.17}$$

$$u_{xx} = 0 \quad \text{in } (0, 1) \setminus I_1 \cup I_2, \tag{3.18}$$

$$u = \psi_i \quad \text{on } I_i, \ i = 1, 2. \tag{3.19}$$

We can now show:

Proposition 3.2. *The function* $e = e(\zeta_1, h_1, \zeta_2, h_2)$ *defined by (3.9) is continuous on the set of points* $\xi = (\zeta_1, h_1, \zeta_2, h_2) \in (r_1, 1 - r_1) \times \mathbb{R} \times (r_2, 1 - r_2) \times \mathbb{R}$.

Proof. As already mentioned in the previous proposition we can relax the assumption (3.4) in the definition of e since $u = u(\xi)$ is perfectly defined when the disks are penetrating each other. Due to the definition (3.9) the continuity of e will follow if we can show that the mapping

$$\xi \mapsto \int_\Omega u(\xi)_x^2(x) \, dx$$

is continuous. ($u(\xi)_x(x)$ denotes the derivative of $u = u(\xi)$ at the point x). But when $\xi' \to \xi$ we have $u(\xi')_x \to u(\xi)_x$ uniformly on each interval where $u(\xi)$ is C^2. It follows that

$$\int_\Omega u(\xi')_x^2 \, dx \to \int_\Omega u(\xi)_x^2 \, dx$$

and the continuity of e is proved. □

We now determine precisely the coincidence sets I_1, I_2. First we have

Proposition 3.3. *In the case of Figure* 15 *it holds that:*

$$x_1 = \zeta_1 - \frac{r_1^2 \zeta_1 + r_1 h_1 \sqrt{h_1^2 + \zeta_1^2 - r_1^2}}{h_1^2 + \zeta_1^2}, \tag{3.20}$$

$$y_1 = \zeta_1 - \frac{r_1}{d^2}\{(h_2 - h_1)\sqrt{d^2 - (r_1 - r_2)^2} - (r_1 - r_2)(\zeta_2 - \zeta_1)\}, \tag{3.21}$$

$$x_2 = \zeta_2 - \frac{r_2}{d^2}\{(h_2 - h_1)\sqrt{d^2 - (r_1 - r_2)^2} - (r_1 - r_2)(\zeta_2 - \zeta_1)\}, \tag{3.22}$$

$$y_2 = \zeta_2 + \frac{r_2^2(1 - \zeta_2) + r_2 h_2 \sqrt{h_2^2 + (1 - \zeta_2)^2 - r_2^2}}{h_2^2 + (1 - \zeta_2)^2} \tag{3.23}$$

where $d = \sqrt{(h_2 - h_1)^2 + (\zeta_2 - \zeta_1)^2}$.

Proof. The first and the last formulae above are obtained exactly as in (2.32), (2.33). To obtain the second formula of (3.21) let us enlarge Figure 15 – see Figure 16

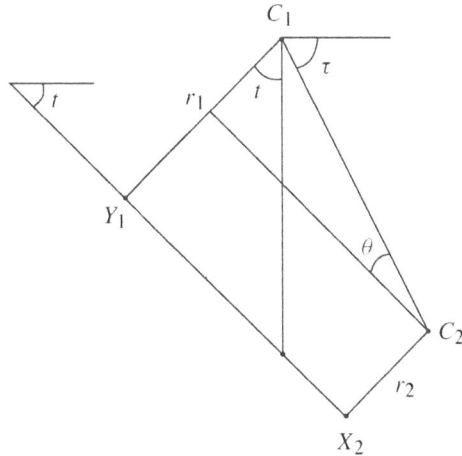

Figure 16. The blow-up of part of Figure 15

We have then clearly

$$y_1 = \zeta_1 - r_1 \sin t, \quad x_2 = \zeta_2 - r_2 \sin t. \tag{3.24}$$

Moreover

$$t = \tau - \theta$$
$$\Rightarrow \quad \sin t = \sin \tau \cos \theta - \sin \theta \cos \tau.$$

Clearly

$$\sin \theta = \frac{r_1 - r_2}{d}, \quad \cos \theta = \frac{\sqrt{d^2 - (r_1 - r_2)^2}}{d}, \tag{3.25}$$

$$\sin \tau = \frac{h_2 - h_1}{d}, \quad \cos \tau = \frac{\zeta_2 - \zeta_1}{d}. \tag{3.26}$$

The formulae (3.21), (3.22) follow.   □

The coincidence set I_1 (see (3.16)) might be not connected when $r_2 < r_1$ and when the small disk is located below the large one – see Figure 17. Then one can show:

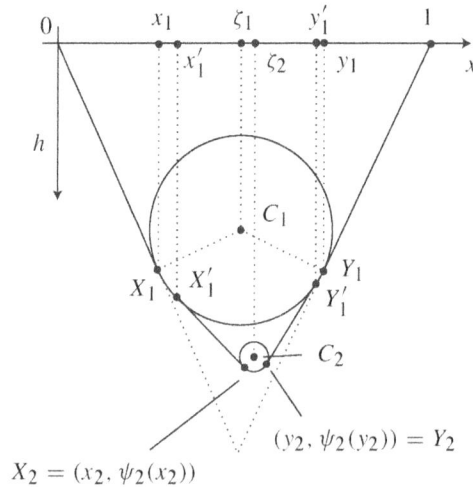

Figure 17. The case of the little disk below the large one

Proposition 3.4. *Assume that we are in the case of Figure* 17. *It holds:*

$$x_1 = \zeta_1 - \frac{r_1^2 \zeta_1 + r_1 h_1 \sqrt{h_1^2 + \zeta_1^2 - r_1^2}}{h_1^2 + \zeta_1^2}, \tag{3.27}$$

$$y_1 = \zeta_1 + \frac{r_1^2(1 - \zeta_1) + r_1 h_1 \sqrt{h_1^2 + (1 - \zeta_1)^2 - r_1^2}}{h_1^2 + (1 - \zeta_1)^2}, \tag{3.28}$$

$$x_1' = \zeta_1 - \frac{r_1}{d^2}\{(h_2 - h_1)(r_1 - r_2) + (\zeta_2 - \zeta_1)\sqrt{d^2 - (r_1 - r_2)^2}\}, \tag{3.29}$$

$$y_1' = \zeta_1 - \frac{r_1}{d^2}\{(\zeta_2 - \zeta_1)\sqrt{d^2 - (r_1 - r_2)^2} - (h_2 - h_1)(r_1 - r_2)\}, \tag{3.30}$$

$$x_2 = \zeta_2 - \frac{r_2}{d^2}\{(h_2 - h_1)(r_1 - r_2) + (\zeta_2 - \zeta_1)\sqrt{d^2 - (r_1 - r_2)^2}\}, \tag{3.31}$$

$$y_2 = \zeta_2 - \frac{r_2}{d^2}\{(\zeta_2 - \zeta_1)\sqrt{d^2 - (r_1 - r_2)^2} - (h_2 - h_1)(r_1 - r_2)\}. \tag{3.32}$$

Proof. It is clear that the formulae (3.27), (3.28) are just (3.20) and (3.23). To prove the other equalities let us enlarge the bottom part of Figure 17 – see Figure 18. As before we have

$$x_1' = \zeta_1 - r_1 \sin t, \quad x_2 = \zeta_2 - r_2 \sin t$$

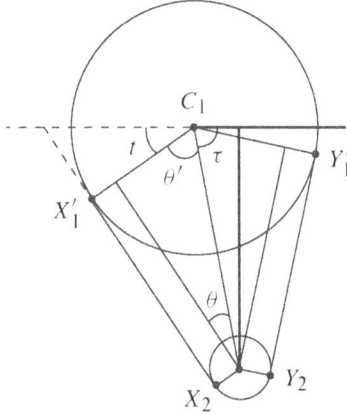

Figure 18. The blow-up of Figure 17

with $\pi = t + \tau + \frac{\pi}{2} - \theta$ i.e. $t = \theta - \tau + \frac{\pi}{2}$. So we have

$$\sin t = \cos(\theta - \tau) = \cos\theta \cos t + \sin\theta \sin\tau$$

with $\sin\theta$, $\cos\theta$, $\sin\tau$, $\cos\tau$ given by (3.25), (3.26). The formulae (3.29), (3.31) follow. The last formulae are obtained the same way. ☐

We can now turn to the differentiability of e. We have first:

Proposition 3.5. *Let* $\xi = (\zeta_1, h_1, \zeta_2, h_2)$ *be a point of* $\mathbb{R}^4$ *such that the configuration is the one of Figure* 15. *Then e is differentiable at* ξ *and it holds:*

$$\frac{\partial e}{\partial \zeta_1}(\xi) = \frac{1}{2}\{\psi_{1,x}^2(y_1) - \psi_{1,x}^2(x_1)\}, \tag{3.33}$$

$$\frac{\partial e}{\partial \zeta_2}(\xi) = \frac{1}{2}\{\psi_{2,x}^2(y_2) - \psi_{2,x}^2(x_2)\}, \tag{3.34}$$

$$\frac{\partial e}{\partial h_1}(\xi) = \psi_{1,x}(x_1) - \psi_{1,x}(y_1) - G_1, \tag{3.35}$$

$$\frac{\partial e}{\partial h_2}(\xi) = \psi_{2,x}(x_2) - \psi_{2,x}(y_2) - G_2, \tag{3.36}$$

where x_i, y_i *are given by* (3.20)–(3.23).

Proof. 1. *Formal proof.* Let us for instance prove (3.35). Differentiating under the integral sum, it comes – see (3.9)

$$\frac{\partial e}{\partial h_1}(\xi) = \int_\Omega u_x u_{xh_1}\, dx - G_1.$$

Integrating by parts we derive easily (see (3.18), (3.19))

$$\frac{\partial e}{\partial h_1}(\xi) = -\int_\Omega u_{xx} u_{h_1}\, dx - G_1$$

$$= -\int_{I_1} \psi_{1,xx}\psi_{1,h_1}\, dx - \int_{I_2} \psi_{2,xx}\psi_{2,h_1}\, dx - G_1$$

$$= -\int_{x_1}^{y_1} \psi_{1,xx}\, dx - G_1$$

and the result follows. The other cases can be treated similarly.

2. *Rigorous proof.* We do it for (3.33). With the notation of Figure 15 we have

$$e = \frac{1}{2}\psi_{1,x}^2(x_1)x_1 + \frac{1}{2}\int_{x_1}^{y_1} \psi_{1,x}^2(x)\, dx - \frac{1}{2}\psi_{1,x}^2(y_1)y_1 + \frac{1}{2}\psi_{2,x}^2(x_2)x_2$$

$$+ \frac{1}{2}\int_{x_2}^{y_2} \psi_{2,x}^2(x)\, dx + \frac{1}{2}(1 - y_2)\psi_{2,x}^2(y_2) - G_1 h_1 - G_2 h_2.$$

Due to Proposition 3.3 the x_i, y_i are differentiable. Thus, taking the derivative in the direction ζ_1 – it is easy to see that every differentiation we will make is justified – we obtain:

$$\frac{\partial e}{\partial \zeta_1}(\xi) = \frac{1}{2}\psi_{1,x}^2(x_1)x_{1,\zeta_1} + x_1\psi_{1,x}(x_1)(\psi_{1,x}(x_1))_{\zeta_1}$$

$$+ \frac{1}{2}\psi_{1,x}^2(y_1)y_{1,\zeta_1} - \frac{1}{2}\psi_{1,x}^2(x_1)x_{1,\zeta_1} +$$

$$+ \int_{x_1}^{y_1} \psi_{1,x}\psi_{1,x\zeta_1}\,dx - \frac{1}{2}\psi_{1,x}^2(y_1)y_{1,\zeta_1} - y_1\psi_{1,x}(y_1)(\psi_{1,x}(y_1))_{\zeta_1}$$

$$+ \frac{1}{2}\psi_{2,x}^2(x_2)x_{2,\zeta_1} + x_2\psi_{2,x}(x_2)(\psi_{2,x}(x_2))_{\zeta_1}$$

$$+ \frac{1}{2}\psi_{2,x}^2(y_2)y_{2,\zeta_1} - \frac{1}{2}\psi_{2,x}^2(x_2)x_{2,\zeta_1}$$

$$+ \int_{x_2}^{y_2} \psi_{2,x}\psi_{2,x\zeta_1}\,dx + (1 - y_2)\psi_{2,x}(y_2)(\psi_{2,x}(y_2))_{\zeta_1} - \frac{1}{2}\psi_{2,x}^2(y_2)y_{2,\zeta_1}.$$

Using the cancellations it comes (note that ψ_2, y_2 are independent of ζ_1)

$$\frac{\partial e}{\partial \zeta_1}(\xi) = x_1\psi_{1,x}(x_1)(\psi_{1,x}(x_1))_{\zeta_1}$$

$$+ \int_{x_1}^{y_1} \psi_{1,x}\psi_{1,x\zeta_1}\,dx - y_1\psi_{1,x}(y_1)(\psi_{1,x}(y_1))_{\zeta_1} \qquad (3.37)$$

$$+ x_2\psi_{2,x}(x_2))(\psi_{2,x}(x_2))_{\zeta_1}.$$

Since $\psi_{1,x}(y_1) = \psi_{2,x}(x_2)$ we have

$$\psi_2(x_2) - \psi_1(y_1) = \psi_{2,x}(x_2)x_2 - \psi_{1,x}(y_1)y_1.$$

Differentiating in ζ_1 we get

$$\psi_{2,x}(x_2)x_{2,\zeta_1} - \psi_{1,\zeta_1}(y_1) - \psi_{1,x}(y_1)y_{1,\zeta_1}$$
$$= \psi_{2,x}(x_2)x_{2,\zeta_1} + (\psi_{2,x}(x_2))_{\zeta_1}x_2 - (\psi_{1,x}(y_1))_{\zeta_1}y_1 - \psi_{1,x}(y_1)y_{1,\zeta_1}.$$

Thus we obtain

$$-\psi_{1,\zeta_1}(y_1) = (\psi_{2,x}(x_2))_{\zeta_1}x_2 - (\psi_{1,x}(y_1))_{\zeta_1}y_1.$$

Since $\psi_{2,x}(x_2) = \psi_{1,x}(y_1)$, $\psi_{1,x} = -\psi_{1,\zeta_1}$ it follows from (3.37) that

$$\frac{\partial e}{\partial \zeta_1}(\xi) = x_1\psi_{1,x}(x_1)(\psi_{1,x}(x_1))_{\zeta_1} - \int_{x_1}^{y_1} \psi_{1,x}\psi_{1,xx}\,dx + \psi_{1,x}(y_1)^2. \qquad (3.38)$$

Using now the equality

$$\psi_1(x_1) = x_1\psi_{1,x}(x_1)$$

and differentiating we obtain

$$\psi_{1,\zeta_1}(x_1) + \psi_{1,x}(x_1)x_{1,\zeta_1} = \psi_{1,x}(x_1)x_{1,\zeta_1} + x_1(\psi_{1,x}(x_1))_{\zeta_1}$$

i.e. $-\psi_{1,x}(x_1) = x_1(\psi_{1,x}(x_1))_{\zeta_1}$.
   Reporting in (3.38) we obtain

$$\frac{\partial e}{\partial \zeta_1}(\xi) = -\psi_{1,x}(x_1)^2 - \frac{1}{2}\int_{x_1}^{y_1} (\psi_{1,x}^2)_x\,dx + \psi_{1,x}(y_1)^2$$

and the result follows by integration. Note that the differentiability follows since the partial derivatives that we obtain are clearly continuous. The other cases can be handled the same way. □

In the case of Figure 17 one has

Proposition 3.6. *Let $\xi = (\xi_1, h_1, \xi_2, h_2)$ be a point of $\mathbb{R}^4$ such that the configuration is the one of Figure* 17. *Then e is differentiable at ξ and it holds:*

$$\frac{\partial e}{\partial \zeta_1}(\xi) = \frac{1}{2}\{\psi_{1,x}^2(x_1') - \psi_{1,x}^2(x_1) + \psi_{1,x}^2(y_1) - \psi_{1,x}^2(y_1')\}, \tag{3.39}$$

$$\frac{\partial e}{\partial \zeta_2}(\xi) = \frac{1}{2}\{\psi_{2,x}^2(y_2) - \psi_{2,x}^2(x_2)\}, \tag{3.40}$$

$$\frac{\partial e}{\partial h_1}(\xi) = \psi_{1,x}(x_1) - \psi_{1,x}(x_1') + \psi_{1,x}(y_1') - \psi_{1,x}(y_1) - G_1, \tag{3.41}$$

$$\frac{\partial e}{\partial h_2}(\xi) = \psi_{2,x}(x_2) - \psi_{2,x}(y_2) - G_2. \tag{3.42}$$

Proof. It is identical to that of Proposition 3.5 and we leave the details to the reader. □

Remark 3.3. In the case of Figure 15 and when $y_2 \to x_2$ one has $\frac{\partial e}{\partial \zeta_2}(\xi) \to 0$, $\frac{\partial e}{\partial h_2}(\xi) \to -G_2$. This corresponds to the case where the second disk does not touch the wire. Similarly when one passes from Figure 18 to Figure 15 the derivatives change smoothly in such a way that the function e is of class C^1 on its domain of definition – i.e. for every ξ such that $\xi \in (r_1, 1 - r_1) \times \mathbb{R} \times (r_2, 1 - r_2) \times \mathbb{R}$.

3.3 Existence result

In order to show that the infimum (3.7) is achieved we will restrict the problem on a compact set by showing that only some configurations has to be considered in the minimization process (3.7). So, let us show

Theorem 3.1. *Under the assumptions* (3.3), (3.11), (3.12) *there exists* (u, C_1, C_2) *achieving the infimum* (3.7).

Proof. We split our arguments in several steps.

Step 1: We can assume $h_1 > -r_1$, or $h_2 > -r_2$.
     If not we have $h_1 \leq -r_1$, $h_2 \leq -r_2$ – i.e. both disks are located above the rest position of the wire and $u = u(\xi) = 0$. Thus one has

$$e = -G_1 h_1 - G_2 h_2 \geq G_1 r_1 + G_2 r_2. \tag{3.43}$$

If one considers only the disk D_1 on the wire it will take an equilibrium position with $h_1 > -r_1$ and an energy $< G_1 r_1$ (see for instance Section 2). Thus for D_1 in this position and D_2 at the level $h_2 = -r_2$ – i.e. in such a way that it does not touch the membrane which is possible by (3.12) – the energy e of the configuration will be

$$e < G_1 r_1 + G_2 r_2$$

which contradicts (3.43).

Step 2: We can assume or consider only the situation $h_1 > -r_1$, $h_2 > -r_2$.

Indeed from step 1 we can assume $h_i > -r_i$ for $i = 1$ or 2. Assume that for $j \neq i$ it holds $h_j \leq -r_j$. Due to (3.12)we can move D_j on one side or another of D_i until it

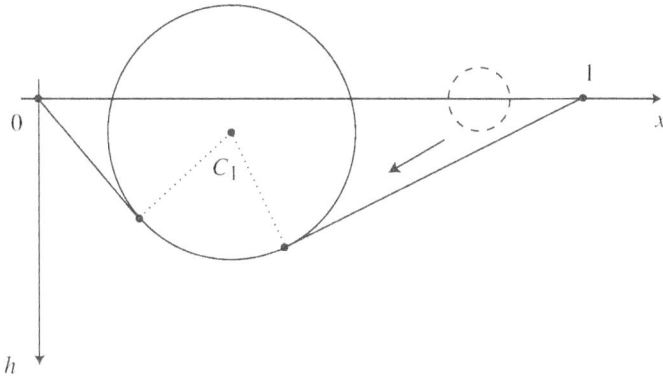

Figure 19. The two disks have to touch the wire

touches the wire and thus is such that $h_j > -r_j$. This process will lower the energy since u will be unchanged but h_j will increase. We can go down until the two disks touch each other (see Figure 19).

Step 3: We can assume $-r_i \leq h_i \leq H$ where H is some constant.

Indeed first we know that we can assume $h_i > -r_i$, $i = 1, 2$. Then, by the Poincaré Inequality there exists a constant c such that

$$e = \frac{1}{2} \int_{\Omega} u_x^2 \, dx - G_1 h_1 - G_2 h_2 \geq \frac{c}{2} \int_{\Omega} u^2 \, dx - G_1 h_1 - G_2 h_2.$$

Choosing $h_i \geq 0$ will cause u to be greater than h_i – see (3.5) – on $B(\zeta_i, r_i)$. In other words we will have

$$e \geq \frac{c}{2} \pi (r_1^2 h_1^2 + r_2^2 h_2^2) - G_1 h_1 - G_2 h_2.$$

Selecting $\|h\|_\infty = Max\{|h_1|, |h_2|\}$ large enough we then force e to be larger than the infimum (3.7) and a contradiction. One can thus assume $h_i \leq H$, $i = 1, 2$ where H is some constant.

Step 4: Analysis of the configuration of Figure 15.

Suppose that $h_i > -r_i$, $i = 1, 2$. Suppose that one of the disks is not touching the wire. Then moving it down will cause a decrease in its energy – see step 2. We can thus assume that both disks are touching the wire. Then we would like to show that in a case like in Figure 15 it is not necessary to consider all the configurations but only the ones where the disks are in contact – i.e. the ones for which the constraint (3.4) is saturated. Indeed consider a case where

$$(\zeta_1 - \zeta_2)^2 + (h_1 - h_2)^2 > (r_1 + r_2)^2. \tag{3.44}$$

We claim that $\nabla e = \left(\frac{\partial e}{\partial \zeta_1}, \frac{\partial e}{\partial h_1}, \frac{\partial e}{\partial \zeta_2}, \frac{\partial e}{\partial h_2} \right) \neq 0$. Indeed, if not, from (3.33), (3.34) we derive

$$\psi_{1,x}^2(y_1) = \psi_{1,x}^2(x_1) = \psi_{2,x}^2(x_2) = \psi_{2,x}^2(y_2).$$

By (3.35), (3.36) – recall that $\psi_{1,x}(y_1) = \psi_{2,x}(x_2)$ – this forces

$$\psi_{1,x}(x_1) = -\psi_{2,x}(y_2) = \frac{G_1 + G_2}{2}.$$

But then $\psi_{1,x}(y_1) = \psi_{2,x}(x_2) = \pm\frac{G_1+G_2}{2}$ which renders $\frac{\partial e}{\partial h_1} = 0$ or $\frac{\partial e}{\partial h_2} = 0$ impossible. Thus one of the derivatives of e is not equal to 0 and moving one disk around will cause a decrease in energy. One can thus assume that (3.4) is saturated. Due to the symmetry of the problem we can assume without loss of generality that

$$\zeta_1 \leq \zeta_2. \tag{3.45}$$

So, consider a configuration of the type of Figure 15 where the two disks are in contact. Suppose that h_1, h_2 are maintained fixed and move the two disks horizontally in order to search the minimum of energy in this situation – i.e. assume

$$\zeta_2 = \zeta_1 + \text{cst} = \zeta_1 + \gamma. \tag{3.46}$$

Due to (3.33), (3.34) it holds that

$$\frac{d}{d\zeta_1} e(\zeta_1, h_1, \zeta_1 + \gamma, h_2) = \left(\frac{\partial e}{\partial \zeta_1} + \frac{\partial e}{\partial \zeta_2} \right)(\zeta_1, h_1, \zeta_1 + \gamma, h_2)$$

$$= \frac{1}{2}\{\psi_{2,x}^2(y_2) - \psi_{1,x}^2(x_1)\}.$$

Thus, the minimum will be achieved for

$$-\psi_{2,x}(y_2) = \psi_{1,x}(x_1).$$

This imposes that the two disks are tangent to the lines with slope

$$\lambda = \psi_{1,x}(x_1) = -\psi_{2,x}(y_2)$$

(see Figure 20). Thus it holds that

$$\zeta_1 \leq \frac{1}{2} \leq \zeta_2. \tag{3.47}$$

We obtain

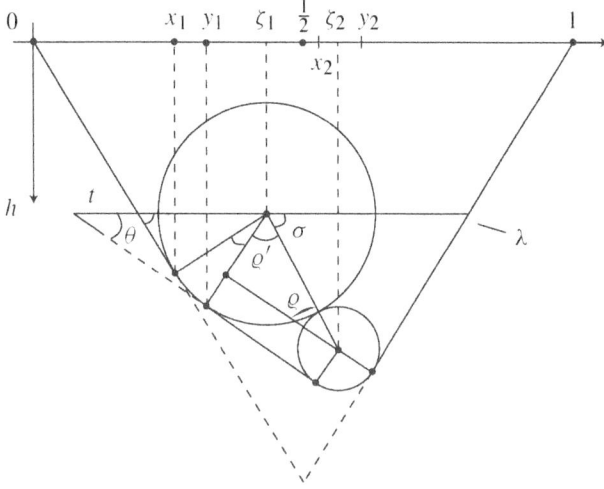

Figure 20

$$\zeta_1 = \zeta_2 + \zeta_1 - \zeta_2 \geq \frac{1}{2} - (r_1 + r_2) \geq \delta + r_1,$$

$$\zeta_2 = \zeta_1 + \zeta_2 - \zeta_1 \leq \frac{1}{2} + (r_1 + r_2) \leq 1 - r_2 - \delta$$

where $\delta = \frac{1}{2} - (2r_1 + r_2) > 0$ by (3.11), (3.12). So, we can assume in this case due to step 3 that $\zeta = (\zeta_1, h_1, \zeta_2, h_2)$ belongs to a compact of $\mathbb{R}^4$.

Step 5: The case of the configuration of Figure 18.

Of course this configuration has to be considered only when $r_2 < r_1$. Then, arguing as in the preceding step – i.e. assuming h_1, h_2 fixed and ζ_1, ζ_2 such that (3.46) holds we have – see (3.39), (3.40)

$$\frac{d}{d\zeta_1} e(\zeta_1, h_1, \zeta_1 + \gamma, h_2)$$

$$= \left(\frac{\partial e}{\partial \zeta_1} + \frac{\partial e}{\partial \zeta_2} \right)(\zeta_1, h_1, \zeta_1 + \gamma, h_2)$$

$$= \frac{1}{2} \{ \psi_{1,x}^2(x_1') - \psi_{1,x}^2(x_1) + \psi_{1,x}^2(y_1) - \psi_{1,x}^2(y_1') + \psi_{2,x}^2(y_2) - \psi_{2,x}^2(x_2) \}$$

$$= \frac{1}{2} \{ \psi_{1,x}^2(y_1) - \psi_{1,x}^2(x_1) \}.$$

Thus the minimum is achieved in this case for

$$-\psi_{1,x}(y_1) = \psi_{1,x}(x_1)$$

i.e. for $\zeta_1 = \frac{1}{2}$. It holds then that

$$\zeta_2 \in \left[\frac{1}{2} - r_2, \frac{1}{2} + r_2\right] \subset (r_2, 1 - r_2)$$

and we can only consider $\xi = (\zeta_1, h_1, \zeta_2, h_2)$ for ξ belonging to a compact subset of $\mathbb{R}^4$.

Step 6: End of the proof.

One minimizes e, a continuous function, (see Proposition 3.2) on a compact subset of $\mathbb{R}^4$. Thus the minimum is achieved. This completes the proof.   ☐

3.4 Research of the minima of energy

We consider first the case where the disks are like on Figure 20. We have seen there that a minimum can be reached only when

$$\psi_{1,x}(x_1) = -\psi_{2,x}(y_2) = \lambda.$$

Now, letting ζ_1, ζ_2 fixed and moving the disks vertically we see that a minimum can be achieved only when

$$\frac{\partial e}{\partial h_1} + \frac{\partial e}{\partial h_2} = 2\lambda - G_1 + G_2 = 0 \quad \Longleftrightarrow \quad \lambda = (G_1 + G_2)/2 = G. \qquad (3.48)$$

In what follows we will assume that (3.48) holds so that $G = (G_1 + G_2)/2$ is the slope of the two tangents (in absolute value) issued from the end points of the wire. Then, see Figure 20, the disks being tangent to these straight lines their position is uniquely determined by the parameter

$$\alpha = \psi_{1,x}(y_1) = \psi_{2,x}(x_2). \qquad (3.49)$$

Let us then compute the energy of the configuration in terms of α. We will set

$$T = \tan \varrho \qquad (3.50)$$

(see Figure 20). Then we have

$$\cos \varrho' = \frac{r_1 - r_2}{r_1 + r_2}, \quad \sin \varrho' = \left\{1 - \left(\frac{r_1 - r_2}{r_1 + r_2}\right)^2\right\}^{1/2} = \frac{2\sqrt{r_1 r_2}}{r_1 + r_2}. \qquad (3.51)$$

Thus,

$$T = \tan \varrho = \frac{1}{\tan \varrho'} = (r_1 - r_2)/2\sqrt{r_1 r_2}. \qquad (3.52)$$

Then, we can show

Proposition 3.7. *Assume that* $T \leq 1/G$. *Then, in the case of Figure 20, the energy* $e = \tilde{e}(\alpha)$ *of the configuration is given by*

$$\tilde{e}(\alpha) = \sqrt{r_1 r_2} \left\{ -T \ln[\sqrt{1 + \alpha^2} + \alpha] + (\alpha + G_1 - G_2) \frac{T + \alpha}{\sqrt{1 + \alpha^2}} + G^2 \frac{1 - \alpha T}{\sqrt{1 + \alpha^2}} \right\}$$

$$+ \frac{r_1 + r_2}{2} \ln[\sqrt{1 + G^2} + G] - \frac{G^2}{2} + \frac{r_1 + r_2}{2} G\sqrt{1 + G^2}. \qquad (3.53)$$

Remark 3.4. For $T > 1/G$ i.e. $\tan \varrho' < G$, the small disk can fit under the large one.

Proof of Proposition 3.7. Using the notation of Figure 20 we have

$$e = \tilde{e}(\alpha) = \frac{1}{2} \int_0^1 u_x^2 \, dx - G_1 h_1 - G_2 h_2$$

$$= \frac{1}{2} G^2 x_1 + \frac{1}{2} \int_{x_1}^{y_1} \psi_{1,x}^2(s) \, ds + \frac{1}{2} \alpha^2 (x_2 - y_1) \qquad (3.54)$$

$$+ \frac{1}{2} \int_{x_2}^{y_2} \psi_{2,x}^2(x) \, dx + \frac{1}{2} G^2 (1 - y_2) - G_1 h_1 - G_2 h_2.$$

We have different quantities to compute. First

$$x_1 = \zeta_1 - r_1 \sin \theta \quad \text{with } \tan \theta = G \implies \cos \theta = \frac{1}{\sqrt{1 + G^2}}, \quad \sin \theta = \frac{G}{\sqrt{1 + G^2}}$$

$$\implies x_1 = \zeta_1 - \frac{r_1 G}{\sqrt{1 + G^2}}. \qquad (3.55)$$

This allows us to get h_1 given by

$$h_1 = Gx_1 - r_1 \cos \theta = G\zeta_1 - r_1\sqrt{1 + G^2}. \qquad (3.56)$$

Since $\tan t = \alpha$ we have $\sin t = \alpha/\sqrt{1 + \alpha^2}$ and

$$y_1 = \zeta_1 - r_1 \sin t = \zeta_1 - \frac{r_1 \alpha}{\sqrt{1 + \alpha^2}}, \qquad (3.57)$$

$$x_2 = \zeta_2 - r_2 \sin t = \zeta_2 - \frac{r_2 \alpha}{\sqrt{1 + \alpha^2}}, \qquad (3.58)$$

$$y_2 = \zeta_2 + r_2 \sin \theta = \zeta_2 + \frac{r_2 G}{\sqrt{1 + G^2}}. \qquad (3.59)$$

Finally, for h_2 we get

$$h_2 = (1 - y_2)G - r_2 \cos \theta = -\zeta_2 G + G - r_2\sqrt{1 + G^2}. \qquad (3.60)$$

We come back now to (3.54). Due to (3.5) we have

$$\psi_{i,x} = -(x - \zeta_i)/\sqrt{r_i^2 - (x - \zeta_i)^2}, \quad i = 1, 2$$

and thus

$$
\psi_{i,x}^2 = (x - \zeta_i)^2 / \{r_i^2 - (x - \zeta_i)^2\}
$$

$$
= \frac{r_i^2}{r_i^2 - (x - \zeta_i)^2} - 1 = \frac{1}{1 - \left(\frac{x - \zeta_i}{r_i}\right)^2} - 1 \quad , i = 1, 2. \tag{3.61}
$$

Thus, for $i = 1, 2$, a simple computation gives

$$
\frac{1}{2} \int_{x_i}^{y_i} \psi_{i,x}^2 (x)\, dx = -\frac{1}{2}(y_i - x_i) + \frac{1}{2} \int_{x_i}^{y_i} \frac{dx}{1 - \left(\frac{x - \zeta_i}{r_i}\right)^2}
$$

$$
= -\frac{1}{2}(y_i - x_i) + \frac{r_i}{2} \int_{(x_i - \zeta_i)/r_i}^{(y_i - \zeta_i)/r_i} dy/(1 - y^2) \tag{3.62}
$$

$$
= -\frac{1}{2}(y_i - x_i) + \frac{r_i}{4} \left\{ \ln \frac{r_i + (y_i - \zeta_i)}{r_i - (y_i - \zeta_i)} - \ln \frac{r_i + (x_i - \zeta_i)}{r_i - (x_i - \zeta_i)} \right\}.
$$

Using now (3.55), (3.57)–(3.59) we obtain

$$
\frac{1}{2} \int_{x_1}^{y_1} \psi_{1,x}^2 (x)\, dx = \frac{1}{2} \left(\frac{r_1 \alpha}{\sqrt{1 + \alpha^2}} - \frac{r_1 G}{\sqrt{1 + G^2}} \right)
$$

$$
+ \frac{r_1}{4} \left\{ \ln \frac{\sqrt{1 + \alpha^2} - \alpha}{\sqrt{1 + \alpha^2} + \alpha} - \ln \frac{\sqrt{1 + G^2} - G}{\sqrt{1 + G^2} + G} \right\}, \tag{3.63}
$$

and

$$
\frac{1}{2} \int_{x_2}^{y_2} \psi_{2,x}^2 (x)\, dx = \frac{1}{2} \left(-\frac{r_2 \alpha}{\sqrt{1 - \alpha^2}} - \frac{r_2 G}{\sqrt{1 + G^2}} \right)
$$

$$
+ \frac{r_2}{4} \left\{ \ln \frac{\sqrt{1 + G^2} + G}{\sqrt{1 + G^2} - G} - \ln \frac{\sqrt{1 + \alpha^2} - \alpha}{\sqrt{1 + \alpha^2} + \alpha} \right\}. \tag{3.64}
$$

Noting that

$$
\ln \frac{\sqrt{1 + x^2} - x}{\sqrt{1 + x^2} + x} = \ln \frac{1}{(\sqrt{1 + x^2} + x)^2} = -2 \ln\left[\sqrt{1 + x^2} + x\right],
$$

we derive from (3.63), (3.64), (3.54) that

$$e = \tilde{e}(\alpha) = -\frac{(r_1 - r_2)}{2} \ln\left[\sqrt{1 + \alpha^2} + \alpha\right]$$

$$+ \frac{r_1 + r_2}{2} \ln\left[\sqrt{1 + G^2} + G\right] + \frac{1}{2}(r_1 - r_2)\frac{\alpha}{\sqrt{1 + \alpha^2}}$$

$$- \frac{1}{2}(r_1 + r_2)\frac{G}{\sqrt{1 + G^2}} + \frac{1}{2}G^2\left(\zeta_1 - \frac{r_1 G}{\sqrt{1 + G^2}}\right)$$

$$+ \frac{1}{2}\alpha^2\left(\zeta_2 - \zeta_1 + (r_1 - r_2)\frac{\alpha}{\sqrt{1 + \alpha^2}}\right)$$

$$+ \frac{1}{2}G^2\left(1 - \zeta_2 - \frac{r_2 G}{\sqrt{1 + G^2}}\right) - G_1 h_1 - G_2 h_2$$

$$= -\frac{(r_1 - r_2)}{2} \ln\left[\sqrt{1 + \alpha^2} + \alpha\right] + \frac{r_1 + r_2}{2} \ln\left[\sqrt{1 + G^2} + G\right]$$

$$+ \frac{1}{2}(r_1 - r_2)\alpha\sqrt{1 + \alpha^2}$$

$$+ \frac{1}{2}G^2(\zeta_1 - \zeta_2) + \frac{1}{2}\alpha^2(\zeta_2 - \zeta_1) + \frac{1}{2}G^2$$

$$- \frac{1}{2}(r_1 + r_2)G\sqrt{1 + G^2} - G_1 h_1 - G_2 h_2. \tag{3.65}$$

We turn now to the computation of h_1 and h_2. We have – see Figure 20

$$\sigma = \varrho + t. \tag{3.66}$$

Thus

$$\tan\sigma = \frac{\tan\varrho + \tan t}{1 - \tan\varrho\tan t} = \frac{T + \alpha}{1 - \alpha T}. \tag{3.67}$$

It follows that it holds

$$\cos^2\sigma = \frac{1}{1 + \tan^2\sigma} = \frac{1}{1 + (\frac{T+\alpha}{1-\alpha T})^2} = \frac{(1 - \alpha T)^2}{(1 + \alpha^2)(1 + T^2)} = \frac{4r_1 r_2}{(r_1 + r_2)^2}\frac{(1 - \alpha T)^2}{1 + \alpha^2}$$

(recall (3.52)). Thus it comes

$$\cos\sigma = \frac{2\sqrt{r_1 r_2}}{r_1 + r_2}\frac{1 - \alpha T}{\sqrt{1 + \alpha^2}}, \quad \sin\sigma = \tan\sigma\cos\sigma = \frac{2\sqrt{r_1 r_2}}{r_1 + r_2}\frac{T + \alpha}{\sqrt{1 + \alpha^2}}. \tag{3.68}$$

We derive then

$$h_2 - h_1 = (r_1 + r_2)\sin\sigma = 2\sqrt{r_1 r_2}\frac{T + \alpha}{\sqrt{1 + \alpha^2}} \tag{3.69}$$

and

$$\zeta_2 - \zeta_1 = (r_1 + r_2)\cos\sigma = 2\sqrt{r_1 r_2}\frac{1 - \alpha T}{\sqrt{1 + \alpha^2}}. \tag{3.70}$$

From (3.56), (3.60) we have also

$$h_1 + h_2 = -G(\zeta_2 - \zeta_1) + G - (r_1 + r_2)\sqrt{1 + G^2}$$

$$= -G2\sqrt{r_1 r_2}\frac{1 - \alpha T}{\sqrt{1 + \alpha^2}} + G - (r_1 + r_2)\sqrt{1 + G^2}. \tag{3.71}$$

Combining (3.69) and (3.71) we get

$$h_1 = -\sqrt{r_1 r_2}\frac{T + \alpha}{\sqrt{1 + \alpha^2}} - \sqrt{r_1 r_2}G\frac{1 - \alpha T}{\sqrt{1 + \alpha^2}} + \frac{G}{2} - \frac{(r_1 + r_2)}{2}\sqrt{1 + G^2}, \tag{3.72}$$

$$h_2 = \sqrt{r_1 r_2}\frac{T + \alpha}{\sqrt{1 + \alpha^2}} - \sqrt{r_1 r_2}G\frac{1 - \alpha T}{\sqrt{1 + \alpha^2}} + \frac{G}{2} - \frac{(r_1 + r_2)}{2}\sqrt{1 + G^2}. \tag{3.73}$$

Using (3.70), (3.72), (3.73) in (3.65) we obtain

$$e = \tilde{e}(\alpha) = -\sqrt{r_1 r_2}T \ln\left[\sqrt{1 + \alpha^2} + \alpha\right] + \frac{r_1 + r_2}{2}\ln\left[\sqrt{1 + G^2} + G\right]$$

$$+ \sqrt{r_1 r_2}T\frac{\alpha(1 + \alpha^2)}{\sqrt{1 + \alpha^2}} - \sqrt{r_1 r_2}G^2\frac{1 - \alpha T}{\sqrt{1 + \alpha^2}} + \sqrt{r_1 r_2}\alpha^2\frac{1 - \alpha T}{\sqrt{1 + \alpha^2}}$$

$$+ \frac{G^2}{2} - \frac{1}{2}(r_1 + r_2)G\sqrt{1 + G^2} + \sqrt{r_1 r_2}(G_1 - G_2)\frac{T + \alpha}{\sqrt{1 + \alpha^2}}$$

$$+ \sqrt{r_1 r_2}2G^2\frac{1 - \alpha T}{\sqrt{1 + \alpha^2}} - G^2 + G(r_1 + r_2)\sqrt{1 + G^2}$$

$$= -\sqrt{r_1 r_2}T \ln\left[\sqrt{1 + \alpha^2} + \alpha\right] + \frac{r_1 + r_2}{2}\ln\left[\sqrt{1 + G^2} + G\right]$$

$$+ \sqrt{r_1 r_2}G^2\frac{1 - \alpha T}{\sqrt{1 + \alpha^2}} + \sqrt{r_1 r_2}(\alpha + G_1 - G_2)\frac{T + \alpha}{\sqrt{1 + \alpha^2}}$$

$$- \frac{G^2}{2} + \frac{r_1 + r_2}{2}G\sqrt{1 + G^2}$$

which completes the proof of (3.53).   □

We are now able to differentiate the energy to get:

Proposition 3.8. *Assume that $T \leq 1/G$, then it holds that*

$$\tilde{e}'(\alpha) = \frac{\sqrt{r_1 r_2}}{(1 + \alpha^2)^{3/2}}\{\alpha^3 - T\alpha^2 + \alpha[2 - G^2 - T(G_1 - G_2)] - G^2 T + G_1 - G_2\}. \tag{3.74}$$

Proof. From (3.53) we derive

$$\tilde{e}'(\alpha) = \sqrt{r_1 r_2}\left\{-\frac{T}{\sqrt{1 + \alpha^2} + \alpha}\left(\frac{\alpha}{\sqrt{1 + \alpha^2}} + 1\right) + \frac{T + \alpha}{\sqrt{1 + \alpha^2}} + \right.$$

$$+ (\alpha + G_1 - G_2) \frac{\sqrt{1 + \alpha^2} - (T + \alpha)\alpha / \sqrt{1 + \alpha^2}}{(1 + \alpha^2)}$$

$$+ G^2 \frac{-T\sqrt{1 + \alpha^2} - (1 - \alpha T)\alpha / \sqrt{1 + \alpha^2}}{(1 + \alpha^2)} \Bigg\}$$

$$= \sqrt{r_1 r_2} \Bigg\{ -\frac{T}{\sqrt{1 + \alpha^2}} + \frac{T + \alpha}{\sqrt{1 + \alpha^2}} + (\alpha + G_1 - G_2) \frac{1 - T\alpha}{(1 + \alpha^2)^{3/2}}$$

$$- G^2 \frac{(T + \alpha)}{(1 + \alpha^2)^{3/2}} \Bigg\}$$

$$= \frac{\sqrt{r_1 r_2}}{(1 + \alpha^2)^{3/2}} \{ \alpha^3 - T\alpha^2 + \alpha[2 - G^2 - T(G_1 - G_2)] - G^2 T + G_1 - G_2 \}.$$

This completes the proof of the proposition. □

In what follows we will set

$$P(\alpha) = \alpha^3 - T\alpha^2 + \alpha[2 - G^2 - T(G_1 - G_2)] - G^2 T + G_1 - G_2. \qquad (3.75)$$

By simple computation we have

$$P(-G) = -G^3 - TG^2 - G[2 - G^2 - T(G_1 - G_2)] - G^2 T + G_1 - G_2$$
$$= -2G(1 + GT) + (G_1 - G_2)(1 + GT) = -2G_2(1 + GT) < 0, \qquad (3.76)$$

and also

$$P(G) = G^3 - TG^2 + G[2 - G^2 - T(G_1 - G_2)] - G^2 T + G_1 - G_2$$
$$= 2G_1(1 - GT). \qquad (3.77)$$

3.4.1 The case of two disks of same size ($T = 0$). In this case we have $r_1 = r_2$ and thus, see (3.52), $T = 0$. The polynomial P above becomes

$$P(\alpha) = \alpha^3 + \alpha(2 - G^2) + G_1 - G_2. \qquad (3.78)$$

3.4.1.1. The case of two identical disks ($G_1 = G_2$). This case was studied in [3] by a Lagrange multiplyer method. Here our approach is different. We have

Theorem 3.2. *If $G \leq \sqrt{2}$ then the two disks have a unique equilibrium position (up to exchanging each other) for $\alpha = 0$ – see Figure 21. If $G > \sqrt{2}$ then the two disks*

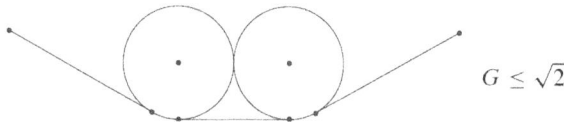

$G \leq \sqrt{2}$

Figure 21. The two disks at the same level

have two equilibrium positions which are symmetric (see Figure 22). If we exchange each other this leads to four possible equilibria.

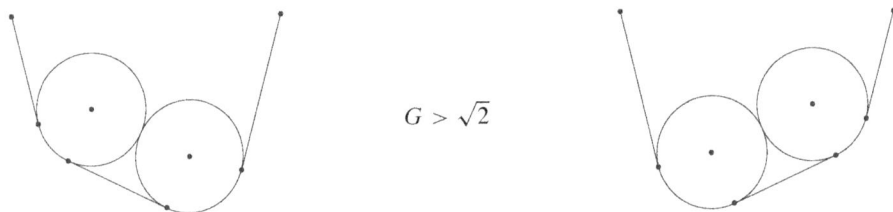

$$G > \sqrt{2}$$

Figure 22. The two disks tilted

Proof. We have by (3.78)

$$P(\alpha) = \alpha[\alpha^2 + 2 - G^2].$$ (3.79)

If $G^2 \leq 2$ then $\alpha = 0$ is the only root of P on $[-G, G]$. Thus since $\tilde{e}$ is decreasing on $(-G, 0)$ increasing on $(0, G)$, $\alpha = 0$ is the only minimum of $\tilde{e}$ on $(-G, G)$ and we are in the case of Figure 21. If now $G^2 > 2$ then it holds that

$$P(\alpha) = \alpha(\alpha - \sqrt{G^2 - 2})(\alpha + \sqrt{G^2 - 2}).$$ (3.80)

$\tilde{e}$ has then two local minima at $\pm\sqrt{G^2 - 2}$ which are also global minima for obvious symmetry reasons. We are in the case of Figure 22. This completes the proof. □

Remark 3.5. The breaking of the symmetry of Figure 21 is somehow a priori unexpected. However, when the weight of the disks is getting higher, – due to the tension of the wire – the equilibrium becomes unstable and the tilting of Figure 22 occurs.

3.4.1.2. The case of different disks in weight ($G_1 \neq G_2$). In this case we can show

Theorem 3.3. *Assume that $G_1 \neq G_2$. Up to a symmetry the equilibrium position of the two disks is unique.*

Proof. By (3.78) we have

$$P'(\alpha) = 3\alpha^2 + 2 - G^2.$$ (3.81)

In the case where $G^2 \leq 2$ then $P'(\alpha) \geq 0$ and P is increasing on $(-G, G)$. Since $P(-G) < 0$, $P(G) > 0$ – see (3.76), (3.77) – P admits a unique 0 on $(-G, G)$ which is the unique minimum of $\tilde{e}$ on $(-G, G)$. This completes the proof in this case. In the case where $G^2 > 2$ then it holds that

$$P'(\alpha) = 3\left(\alpha - \frac{\sqrt{G^2 - 2}}{\sqrt{3}}\right)\left(\alpha + \frac{\sqrt{G^2 - 2}}{\sqrt{3}}\right).$$ (3.82)

Let us set

$$\alpha_\pm = \pm\sqrt{G^2 - 2}/\sqrt{3}. \tag{3.83}$$

It is clear that P is increasing on $(-G, \alpha_-)$, decreasing on (α_-, α_+), increasing on (α_+, G). Thus if

$$P(\alpha_-) \leq 0 \quad \text{or} \quad P(\alpha_+) \geq 0 \tag{3.84}$$

then P changes its sign only once and $\tilde{e}$ has a unique minimum on $(-G, G)$ – recall that $P(-G) < 0$, $P(G) > 0$. This completes the proof in this case. If now

$$P(\alpha_-) > 0 \quad \text{and} \quad P(\alpha_+) < 0 \tag{3.85}$$

then P admits three zeros on $(-G, G)$ and $\tilde{e}$ admits two local minima on $(-G, G)$. However, only one is a global minimum. This is due to the fact that the light disk will always sit above the more heavy one. Let us denote by $\alpha_1, \alpha_2, \alpha_1 < \alpha_- < \alpha_+ < \alpha_2$ the two points of $(-G, G)$ where $\tilde{e}$ achieves its local minimum. Suppose that

$$G_2 < G_1. \tag{3.86}$$

Then, $\tilde{e}$ cannot achieve a global minimum in α_2 where the second disk is in a position lower than the first one. Indeed suppose that

$$e_1 = \tilde{e}(\alpha_2) = \frac{1}{2}\int_\Omega u_x^2 - G_1 h_1 - G_2 h_2.$$

Exchanging the two disks will produce a configuration with an energy

$$e_2 = \frac{1}{2}\int_\Omega u_x^2 - G_2 h_1 - G_1 h_2$$

so that

$$e_2 - e_1 = (h_1 - h_2)(G_1 - G_2) < 0 \quad (h_1 < h_2) \tag{3.87}$$

recall that h is directed downward. Thus, by (3.87), the energy e_2 is smaller than e_1 which renders impossible to have a minimum at α_2. Of course when $G_2 > G_1$ one would prove the same way that $\tilde{e}$ cannot have an absolute minimum in α_1. This completes the proof of the theorem. □

Remark 3.6. One has – see (3.78):

$$P(\alpha_+) = \pm\left\{\frac{(G^2 - 2)^{3/2}}{3\sqrt{3}} - \frac{(G^2 - 2)^{3/2}}{\sqrt{3}}\right\} + G_1 - G_2$$

$$= \pm\left\{-\frac{2}{3}\frac{(G^2 - 2)^{3/2}}{\sqrt{3}}\right\} + G_1 - G_2. \tag{3.88}$$

Thus (3.85) holds if we have

$$-2\left(\frac{G^2-2}{3}\right)^{3/2} < G_1 - G_2 < 2\left(\frac{G^2-2}{3}\right)^{3/2}$$

$$\Longleftrightarrow \qquad |G_1 - G_2| < 2\left(\frac{G^2-2}{3}\right)^{3/2}$$

(3.89)

which can of course occur.

For the case of different disks in size and weight we refer the reader to [11].

3.5 Further problems

In this last section we would like to mention briefly several other issues which have yet to be investigated. To consider the case of two ellipses free to roll on an elastic wire would be probably a nightmare to study and we do not insist on this topic. A natural continuation of the problem of two disks would be the problem of n identical disks – (see [1] for very partial results). Then – in the case of three disks for instance – a new situation can occur: one disk can sit atop of the two others and a special investigation of the energy has to be made in this case (see Figure 23). Of course with more disks

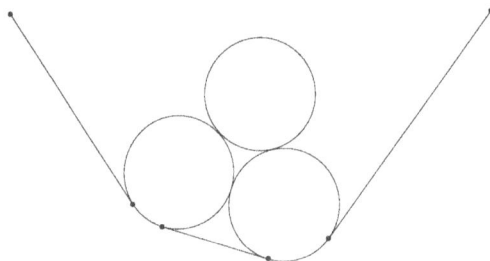

Figure 23. A new situation in the case of three disks

the possible packing of the disks together can become more involved! We let this to the curiosity of the reader.

4 The case of a single ball on an elastic membrane

In this section Ω is a bounded domain of $\mathbb{R}^2$ with a smooth boundary Γ. As we mentioned in our introduction Γ is a horizontal frame on which an elastic membrane is spanned, Ω is the undeformed position of the membrane. Letting roll freely on this membrane a heavy ball of weight G and radius r we are interested in finding the

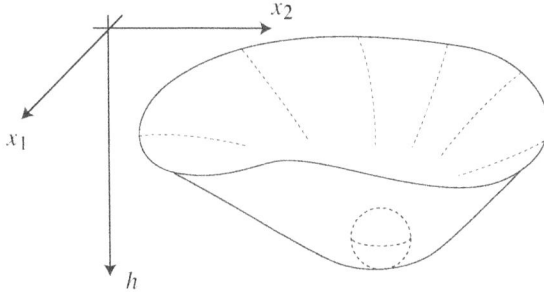

Figure 24. The equilibrium position of a ball on a membrane

equilibrium position reached by the ball. As before in the case of a disk on a wire, the position of the ball is completely determined by the position of its center

$$C = (\zeta, h) = (\zeta_1, \zeta_2, h), \tag{4.1}$$

(recall that the vertical axis is oriented downward). The function describing the bottom part of the ball is now given by – see Figure 4 –

$$\psi = \psi^{\zeta h}(x) = h + \sqrt{r^2 - |x - \zeta|^2}, \quad x \in B(\zeta, r), \tag{4.2}$$

where $B(\zeta, r)$ denotes the open ball in $\mathbb{R}^2$ of center ζ and radius r – i.e.

$$B(\zeta, r) = \{x \in \mathbb{R}^2 \mid |x - \zeta| < r\}. \tag{4.3}$$

$|\cdot|$ is the euclidean norm in $\mathbb{R}^2$ defined by

$$|z| = \{z_1^2 + z_2^2\}^{1/2} \quad \forall z = (z_1, z_2) \in \mathbb{R}^2. \tag{4.4}$$

For a position (ζ, h) of the center of the ball an admissible deformation is an element u of $K = K(\zeta, h)$ defined by

$$K = K(\zeta, h) = \{v \in H_0^1(\Omega) \mid v(x) \geq \psi^{\zeta h}(x) \text{ a.e. } x \in B(\zeta, r)\}. \tag{4.5}$$

It is easy to check that $K(\zeta, h)$ is a closed convex set of $H_0^1(\Omega)$. In order to avoid the ball to be in contact to the frame Γ we will consider only positions of the center such that

$$\text{dist}(\zeta, \Gamma) > r \tag{4.6}$$

where

$$\text{dist}(\zeta, \Gamma) = \inf_{z \in \Gamma} |\zeta - z|. \tag{4.7}$$

If we define

$$\Omega_r = \{\zeta \in \Omega \mid \text{dist}(\zeta, \Gamma) > r\} \tag{4.8}$$

we will choose

$$(\zeta, h) \in \Omega_r \times \mathbb{R} \tag{4.9}$$

and for such a point the energy of an admissible deformation u is given by

$$E(u; \zeta, h) = \frac{1}{2} \int_\Omega |\nabla u(x)|^2 \, dx - Gh. \tag{4.10}$$

(∇u denotes the usual gradient i.e. the vector of components $\frac{\partial u}{\partial x_1}$, $\frac{\partial u}{\partial x_2}$). Then to look for an equilibrium position of the ball reduces to look for an element (u_0, ζ_0, h_0) such that

$$\begin{cases} (\zeta_0, h_0) \in \Omega_r \times \mathbb{R}, \quad u_0 \in K(\zeta_0, h_0), \\ E(u_0; \zeta_0, h_0) \leq E(u; \zeta, h) \quad \forall (\zeta, h) \in \Omega_r \times \mathbb{R}, \quad \forall u \in K(\zeta, h). \end{cases} \tag{4.11}$$

As observed in previous situations, if such a point u_0 does exist then it holds that

$$E(u_0; \zeta_0, h_0) \leq E(u; \zeta_0, h_0) \quad \forall u \in K(\zeta_0, h_0)$$

that is to say u_0 satisfies $u_0 \in K(\zeta_0, h_0)$ and

$$\Longleftrightarrow \quad \begin{aligned} \frac{1}{2} \int_\Omega |\nabla u_0|^2 \, dx - Gh_0 &\leq \frac{1}{2} \int_\Omega |\nabla u|^2 \, dx - Gh_0 \quad \forall u \in K(\zeta_0, h_0), \\ \frac{1}{2} \int_\Omega |\nabla u_0|^2 \, dx &\leq \frac{1}{2} \int_\Omega |\nabla u|^2 \, dx \quad \forall u \in K(\zeta_0, h_0). \end{aligned} \tag{4.12}$$

It is well known that $H_0^1(\Omega)$ is a Hilbert space when equipped with the norm

$$\|\nabla u\|_2 = \left\{ \int_\Omega |\nabla u|^2 \, dx \right\}^{1/2}. \tag{4.13}$$

Thus, by (4.12), u_0 is the projection of 0 on the closed convex set $K(\zeta_0, h_0)$ and also the solution of the variational inequality (see Proposition 2.1)

$$\begin{cases} u_0 \in K(\zeta_0, h_0), \\ \int_\Omega \nabla u_0 \cdot \nabla(v - u_0) \, dx \geq 0 \quad \forall v \in K(\zeta_0, h_0). \end{cases} \tag{4.14}$$

(We denote with a dot the scalar product.) Thus, for $(\zeta, h) \in \Omega_r \times \mathbb{R}$ we can introduce $u = u(\zeta, h)$ the solution to

$$\begin{cases} u \in K(\zeta, h), \\ \int_\Omega \nabla u \cdot \nabla(v - u) \, dx \geq 0 \quad \forall v \in K(\zeta, h). \end{cases} \tag{4.15}$$

(Such a solution exists and is unique – see Proposition 2.1.) For $u = u(\zeta, h)$ we can then define e by

$$e(\zeta, h) = \frac{1}{2} \int_{\Omega} |\nabla u|^2 \, dx - Gh. \tag{4.16}$$

Due to (4.12) it is clear that if (u_0, ζ_0, h_0) is a minimizer to (4.11) then (ζ_0, h_0) is a minimizer of e on $\Omega_r \times \mathbb{R}$. Conversely if (ζ_0, h_0) is a minimizer of e on $\Omega_r \times \mathbb{R}$, then for u_0 solution to (4.14), (u_0, ζ_0, h_0) is a minimizer to (4.11). Thus, we are now reduced to find a point (ζ_0, h_0) such that

$$\begin{cases} (\zeta_0, h_0) \in \Omega_r \times \mathbb{R}, \\ e(\zeta_0, h_0) \le e(\zeta, h) \quad \forall \, (\zeta, h) \in \Omega_r \times \mathbb{R}, \end{cases} \tag{4.17}$$

i.e. we are reduced to minimize e on $\Omega_r \times \mathbb{R}$. It should be noticed that this later set is not compact, thus the problem requires some work in order to be solved.

Remark 4.1. In this note we will restrict ourselves to the Dirichlet integral. One could consider also an energy of the type

$$E(u; \zeta, h) = \frac{1}{2} \int_{\Omega} \sqrt{1 + |\nabla u(x)|^2} \, dx - Gh \tag{4.18}$$

which corresponds to a membrane of a soap film type. Of course in this case in order for the membrane not to brake we have to impose some limit to h. We refer the reader to [4] for details.

4.1 Some remarks on the obstacle problem

In this section we would like to study more deeply the problem (4.15). This is a problem of obstacle type – i.e. the solution is forced to stay above an "obstacle" (see [5], [7], [8], [18]). However, this obstacle problem is not a classical one due to the fact that $\psi = \psi^{\zeta h}$ – the so called obstacle – is not defined on the whole domain Ω and moreover its gradient does not remain bounded. So, it requires some analysis. First we would like to show that the solution u of (4.15) is such that

$$u \in W^{2, \infty}(\Omega). \tag{4.19}$$

(We refer the reader to [13], [17], [18] for notation and results on Sobolev spaces.) For that, let us first prove:

Lemma 4.1 (Monotonicity of the solution with respect to the domain). *Let $\Omega^* \subset \Omega$ be an open subset of Ω containing $B(\zeta, r)$. Let u^* be the solution to*

$$\begin{cases} u^* \in K^* = K^*(\zeta, h) = \{ v \in H_0^1(\Omega^*) \mid u(x) \ge \psi^{\zeta h}(x) \text{ a.e. } x \in B(\zeta, r) \}, \\ \int_{\Omega^*} \nabla u^* \cdot \nabla(v - u^*) \, dx \ge 0 \quad \forall v \in K^*. \end{cases}$$

$$\tag{4.20}$$

Let u be the solution to (4.15). Then if u^* is supposed to be extended by 0 on $\Omega \setminus \Omega^*$ it holds that

$$u^*(x) \leq u(x) \quad a.e.\ x \in \Omega. \tag{4.21}$$

Proof. If $(\cdot)^+$ denotes the positive part of a function we want to show that

$$(u^* - u)^+ = 0. \tag{4.22}$$

For that, we remark first that u and u^* are nonnegative. Let us show it for u, the proof for u^* being similar. We remark that

$$v = u^+ \in K(\zeta, h) \tag{4.23}$$

and thus by (4.15) it holds that

$$\int_\Omega \nabla u \cdot \nabla(u^+ - u)\,dx \geq 0. \tag{4.24}$$

If $u^- = (-u)^+$ denotes the negative part of u we have $u = u^+ - u^-$ and (4.24) leads to

$$0 \leq \int_\Omega \nabla u \cdot \nabla u^-\,dx = -\int_\Omega \nabla u^- \cdot \nabla u^-\,dx, \tag{4.25}$$

(see [9]). It follows that $u^- = 0$ i.e. $u \geq 0$ a.e. in Ω. This is what we wanted to prove first. Next, we notice that

$$v = u^* - (u^* - u)^+ \in H_0^1(\Omega^*). \tag{4.26}$$

(This is due to the fact that $u \geq 0$ on the boundary of Ω^*.) Now

$$v = u^* \text{ or } u,$$

it is clear that $v \in K^*(\zeta, h)$ and from (4.20) we derive

$$\int_\Omega \nabla u^* \cdot \nabla(u^* - u)^+\,dx \leq 0. \tag{4.27}$$

Now since

$$v = u + (u^* - u)^+ = u \text{ or } u^*$$

also, this is a suitable test function for (4.15) and we obtain

$$\int_\Omega -\nabla u \cdot \nabla(u^* - u)^+\,dx \leq 0. \tag{4.28}$$

Summing up (4.27) and (4.28) it comes

$$\int_\Omega \nabla(u^* - u)\nabla(u^* - u)^+\,dx = \int_\Omega |\nabla(u^* - u)^+|^2\,dx \leq 0 \tag{4.29}$$

which implies (4.22) and completes the proof of the lemma.   □

Let us denote by Λ the set

$$\Lambda = \{x \in B(\zeta, r) \mid u(x) = \psi^{\zeta h}(x)\}. \tag{4.30}$$

Λ is the coincidence set of the obstacle problem (4.15). For the time being this is a measurable subset of $B(\zeta, r)$. We are going to show that this is a closed subset of $B(\zeta, r)$. For that, let us denote by R the distance from ζ to Γ the boundary of Ω – i.e.

$$R = \mathrm{dist}(\zeta, \Gamma) = \mathop{\mathrm{Inf}}_{\gamma \in \Gamma} |\zeta - \gamma|. \tag{4.31}$$

Clearly for $\zeta \in \Omega_r$ it holds that

$$R > r. \tag{4.32}$$

Let us set

$$\Omega^* = B(\zeta, R). \tag{4.33}$$

Then we have:

Lemma 4.2. *Suppose that* $\zeta \in \Omega_r$, $h > -r$. *There is a unique* $s \in (0, r)$ *such that*

$$f(s) = \ln \frac{s}{R} + \frac{1}{s^2}\left\{h\sqrt{r^2 - s^2} + r^2 - s^2\right\} = 0. \tag{4.34}$$

Moreover, if Ω^* *is given by (4.33) the solution to (4.20) is given by*

$$u^*(x) = \begin{cases} \psi^{\zeta h}(x) & \text{if } \varrho = |x - \zeta| \le s, \\ \dfrac{-s^2}{\sqrt{r^2 - s^2}}\ln\dfrac{\varrho}{R} & \text{if } \varrho = |x - \zeta| \ge s. \end{cases} \tag{4.35}$$

In particular u^* *is radially symmetric,*

$$u^* \in W^{2, \infty}(\Omega^*) \tag{4.36}$$

and the coincidence set is given by

$$\Lambda^* = \{x \in \Omega^* \mid u^*(x) = \psi^{\zeta h}(x)\} = \bar{B}(\zeta, s) \tag{4.37}$$

where $\bar{B}(\zeta, s)$ *denotes the closed ball of center* ζ *and radius* s.

Proof. If f is defined by (4.34) – i.e. if we set

$$f(\varrho) = \ln \frac{\varrho}{R} + \frac{1}{\varrho^2}\{h\sqrt{r^2 - \varrho^2} + r^2 - \varrho^2\}, \tag{4.38}$$

we have

$$\begin{aligned}
f'(\varrho) &= \frac{1}{\varrho} + \frac{1}{\varrho^2}\left\{\frac{-h\varrho}{\sqrt{r^2 - \varrho^2}} - 2\varrho\right\} - \frac{2}{\varrho^3}\{h\sqrt{r^2 - \varrho^2} + r^2 - \varrho^2\} \\
&= \frac{\varrho^2 - 2r^2}{\varrho^3} - \frac{h}{\varrho\sqrt{r^2 - \varrho^2}} - \frac{2h}{\varrho^3}\sqrt{r^2 - \varrho^2} < 0 \quad \text{for } \varrho \in (0, r).
\end{aligned} \tag{4.39}$$

This shows that f is decreasing. Moreover since

$$\lim_{\varrho \to 0} f(\varrho) = +\infty, \qquad f(r) = \ln \frac{r}{R} < 0$$

it is clear that there exists a unique $s \in (0, r)$ such that (4.34) holds. Now (4.34) can also be written

$$h + \sqrt{r^2 - s^2} = \frac{-s^2}{\sqrt{r^2 - s^2}} \ln \frac{s}{R} \tag{4.40}$$

which shows that u^* given by (4.35) and $\psi^{\zeta h}$ coincide for $\varrho = s$. Moreover, since u^* and ψ are radially symmetric we have with obvious notation

$$\frac{d\psi}{d\varrho}(s) = \frac{-s}{\sqrt{r^2 - s^2}} = \frac{du^*}{d\varrho}(s). \tag{4.41}$$

Thus the function u^* defined by (4.35) is a radially symmetric C^1-function and thus belongs clearly to $W^{2,\infty}(\Omega^*)$. Next, due to (4.39), (4.40) it holds that

$$h + \sqrt{r^2 - \varrho^2} < \frac{-\varrho^2}{\sqrt{r^2 - \varrho^2}} \ln\left(\frac{\varrho}{R}\right) < \frac{-s^2}{\sqrt{r^2 - s^2}} \ln\left(\frac{\varrho}{R}\right) \quad \forall \varrho \in (s, r), \tag{4.42}$$

(the function $\varrho \mapsto \dfrac{\varrho^2}{\sqrt{r^2 - \varrho^2}}$ is clearly increasing) and thus (4.37) holds. To see that u^* given by (4.35) is solution to (4.20) one remarks first that (4.42) implies that

$$u^* \in K^*(\zeta, h). \tag{4.43}$$

Next we remark that u^* is harmonic outside $B(\zeta, s)$. Indeed outside this ball we have if we use the Laplace operator in polar coordinates

$$\Delta u^* = \frac{d^2 u^*}{d\varrho^2} + \frac{1}{\varrho} \frac{du^*}{d\varrho} = \frac{-s^2}{\sqrt{r^2 - s^2}} \left\{ -\frac{1}{\varrho^2} + \frac{1}{\varrho^2} \right\} = 0. \tag{4.44}$$

Thus we have, if we denote simply $B(\zeta, s)$ by B

$$\int_{\Omega^* \backslash B} \nabla u^* \cdot \nabla(v - u^*) \, dx = \int_{\Omega^* \backslash B} \nabla \cdot \{(v - u^*)\nabla u^*\} \, dx$$

$$= -\int_{\partial B} (v - u^*) \frac{\partial u^*}{\partial n} \, d\sigma(x) \quad \forall v \in K^* \tag{4.45}$$

($\frac{\partial u^*}{\partial n}$ denotes the derivative in the direction of the outward unit normal n to ∂B the boundary of B, $d\sigma$ is the surface area on ∂B – recall that v, u^* vanish on the boundary of Ω^* and u^* is a C^1-function). Similarly on B we have

$$\int_B \nabla u^* \cdot \nabla(v - u^*) \, dx = \int_B \nabla \cdot \{(v - u^*)\nabla u^*\} - (v - u^*)\Delta u^* \, dx$$

$$= \int_{\partial B} (v - u^*) \frac{\partial u^*}{\partial n} \, d\sigma(x) - \int_B (v - \psi)\Delta\psi \, dx. \tag{4.46}$$

Summing up (4.45) and (4.46) leads to

$$\int_{\Omega^*} \nabla u^* \cdot \nabla(v - u^*)\,dx = -\int_B (v - \psi)\Delta\psi\,dx. \tag{4.47}$$

Now arguing like in (4.44) we have

$$\Delta\psi = \frac{d^2\psi}{d\varrho^2} + \frac{1}{\varrho}\frac{d\psi}{d\varrho} \tag{4.48}$$

with

$$\frac{d\psi}{d\varrho} = \frac{-\varrho}{\sqrt{r^2 - \varrho^2}}, \quad \frac{d^2\psi}{d\varrho^2} = \left\{ -\sqrt{r^2 - \varrho^2} - \frac{(-\varrho)(-\varrho)}{\sqrt{r^2 - \varrho^2}} \right\}/(r^2 - \varrho^2). \tag{4.49}$$

Thus

$$\frac{d^2\psi}{d\varrho^2} = \frac{-(r^2 - \varrho^2) - \varrho^2}{(r^2 - \varrho^2)^{3/2}} = -\frac{r^2}{(r^2 - \varrho^2)^{3/2}}. \tag{4.50}$$

Combining this with (4.48) we get

$$\Delta\psi = \frac{-r^2}{(r^2 - \varrho^2)^{3/2}} - \frac{1}{\sqrt{r^2 - \varrho^2}} = \frac{\varrho^2 - 2r^2}{(r^2 - \varrho^2)^{3/2}} \leq 0. \tag{4.51}$$

Since for $v \in K^*(\zeta, h)$ we have

$$v - \psi \geq 0 \quad \text{on } B(\zeta, r),$$

we deduce from (4.47), (4.51)

$$\int_{\Omega^*} \nabla u^* \cdot \nabla(v - u^*)\,dx \geq 0 \quad \forall v \in K^*(\zeta, h)$$

and thus u^* is solution to (4.20). This completes the proof of the lemma. □

Let ζ be a point of Ω_r. Let u^* be the function defined by (4.35) that we suppose to be extended by the bottom formula of (4.35) outside of $B(\zeta, R)$. Then, clearly, as seen in the preceding lemma we have

$$u^* \in W^{2,\infty}(\Omega). \tag{4.52}$$

Moreover:

Lemma 4.3. *Let $\zeta \in \Omega_r$, $h > -r$. Let u be the unique solution to (4.15) then u is also the unique solution to the variational inequality*

$$\begin{cases} u \in K_{u^*} = \{v \in H_0^1(\Omega) \mid v(x) \geq u^*(x) \text{ a.e. } x \in \Omega\}, \\ \int_\Omega \nabla u \cdot \nabla(v - u)\,dx \geq 0 \quad \forall v \in K_{u^*}. \end{cases} \tag{4.53}$$

Proof. It is clear that K_{u^*} is a closed convex set of $H_0^1(\Omega)$, so the existence and uniqueness of a solution to (4.53) follows the arguments that we developed before.

Let us, for the time being, denote by u' the solution to (4.53). Since – see (4.43) – $u^* \in K^*(\zeta, h)$ we have clearly $u' \in K(\zeta, h)$ so that by (4.15)

$$\int_\Omega \nabla u \cdot \nabla(u' - u)\, dx \geq 0. \tag{4.54}$$

Next, due to Lemmas 4.1 and 4.2 we have

$$u \geq u^*, \quad \text{i.e.,} \quad u \in K_{u^*}$$

and by (4.53) we derive

$$\int_\Omega \nabla u' \cdot \nabla(u - u')\, dx \geq 0. \tag{4.55}$$

Summing now (4.54) and (4.55) we get

$$\int_\Omega |\nabla(u - u')|^2\, dx \leq 0$$

i.e. $u = u'$. This completes the proof of the lemma. ☐

Then we are now able to prove

Proposition 4.1. *Let $\zeta \in \Omega_r$, $h > -r$. Let u be the solution to (4.15) then it holds*

$$u \in W^{2,\infty}(\Omega). \tag{4.56}$$

The coincidence set

$$\Lambda = \{ x \in B(\zeta, r) \mid u(x) = \psi^{\zeta h}(x) \} \tag{4.57}$$

is a closed subset of $B(\zeta, r)$ such that for some $\varepsilon > 0$

$$\Lambda \subset B(\zeta, r - \varepsilon). \tag{4.58}$$

Moreover, it holds that

$$-\Delta u \geq 0 \quad \text{in } \Omega, \tag{4.59}$$

$$-\Delta u = 0 \quad \text{in } \Omega \setminus \Lambda \tag{4.60}$$

((4.59) *means that Δu is a nonpositive measure in Ω,* (4.60) *means that u is harmonic outside the coincidence set*).

Proof. By Lemma 4.3 u is solution to (4.53) with $u^* \in W^{2,\infty}(\Omega)$. It follows by well known results of regularity – see [5], [7], [8], [19] – that (4.56) holds. Moreover, see (4.42), we have

$$\Lambda \subset B(\zeta, s) \tag{4.61}$$

which is (4.58). The fact that Λ is closed follows from the fact that both u and ψ are continuous. Now let us consider

$$\varphi \in \mathcal{D}(\Omega), \quad \varphi \geq 0, \tag{4.62}$$

where $\mathcal{D}(\Omega)$ denotes the space of infinitely differentiable functions with compact support in Ω. It is clear that $v = u + \varphi \in K(\zeta, h)$ and v is a suitable test function for (4.15). Thus we have

$$\int_\Omega \nabla u \cdot \nabla \varphi \, dx \geq 0 \quad \forall \varphi \in \mathcal{D}(\Omega), \ \varphi \geq 0 \tag{4.63}$$

$$\Longleftrightarrow \qquad \langle -\Delta u, \varphi \rangle \geq 0 \quad \forall \varphi \in \mathcal{D}(\Omega), \ \varphi \geq 0,$$

$\langle \cdot, \cdot \rangle$ denotes the duality bracket between $\mathcal{D}'(\Omega)$ and $\mathcal{D}(\Omega)$, see [9]. This is (4.59). From this and (4.44) we deduce

$$-\Delta(u - u^*) = -\Delta u \geq 0 \quad \text{in } \Omega \setminus B(\zeta, s), \tag{4.64}$$

$$u - u^* \geq 0 \quad \text{in } \Omega \setminus \bar{B}(\zeta, s). \tag{4.65}$$

Thus, by the strong maximum principle – see [17] – we deduce that

$$u > u^* \quad \text{in } \Omega \setminus \bar{B}(\zeta, s). \tag{4.66}$$

Since on $\bar{B}(\zeta, s)$ we have

$$u^* = \psi, \tag{4.67}$$

it follows that

$$\Lambda = \{ x \in B(x, r) \mid u(x) = \psi(x) \} = \{ x \in \Omega \mid u(x) = u^*(x) \}. \tag{4.68}$$

Let then φ be a function such that

$$\varphi \in \mathcal{D}(\Omega \setminus \Lambda). \tag{4.69}$$

The set

$$\Omega \setminus \Lambda = \{ x \in \Omega \mid u(x) > u^*(x) \} \tag{4.70}$$

is of course an open subset. Since φ has compact support for ε small enough it holds that

$$u \pm \varepsilon \varphi \geq u^* \quad \text{in } \Omega. \tag{4.71}$$

Thus, $v = u \pm \varepsilon \varphi$ are both test functions for (4.53) and we obtain

$$\pm \varepsilon \int_\Omega \nabla u \cdot \nabla \varphi \, dx \geq 0 \quad \forall \varphi \in \mathcal{D}(\Omega \setminus \Lambda), \tag{4.72}$$

i.e.,

$$-\Delta u = 0 \quad \text{in } \Omega \setminus \Lambda \tag{4.73}$$

in the distributional sense or almost everywhere by (4.56). This completes the proof of the proposition. $\qquad \square$

4.2 Existence of an equilibrium

As seen previously the minimization problem (4.11) reduces to (4.17) – i.e. to the
minimization of the function $e(\zeta, h)$ defined by (4.16). This is what we would like
to address now. Note that the preceding section provides us with some information
about the function u used in the definition of e. First we would like to show that the
function e is continuous on $\Omega_r \times \mathbb{R}$. This will be a direct consequence of the following
lemma:

Lemma 4.4. *Let* $(\zeta_n, h_n) \in \Omega_r \times \mathbb{R}$ *be a sequence of points such that*

$$(\zeta_n, h_n) \to (\zeta, h) \in \Omega_r \times \mathbb{R}. \tag{4.74}$$

If $u_n = u(\zeta_n, h_n)$, $u = u(\zeta, h)$ *denote the solutions to (4.15) we have*

$$u_n \to u \quad in \ H_0^1(\Omega). \tag{4.75}$$

Proof. Step 1: We claim that there exists n_0 such that

$$\bigcap_{n \geq n_0} K(\zeta_n, h_n) \cap K(\zeta, h) \neq \emptyset \tag{4.76}$$

i.e. there exists a function v belonging to all the $K(\zeta_n, h_n)$, $n \geq n_0$ and to $K(\zeta, h)$.
For that set

$$\delta = \frac{1}{2}\{\operatorname{dist}(\zeta, \Gamma) - r\}. \tag{4.77}$$

Then, for $n \geq n_0$ we have

$$|\zeta_n - \zeta| < \delta \tag{4.78}$$

and thus

$$B(\zeta_n, r) \subset B(\zeta, r + \delta) \quad \forall n \geq n_0. \tag{4.79}$$

Since

$$\psi^{\zeta_n h_n}(x) = h_n + \sqrt{r^2 - |x - \zeta_n|^2} \leq h_n + r \tag{4.80}$$

and since h_n is bounded – since converging – there exists a constant $H > 0$ such that

$$\psi^{\zeta_n h_n}(x) \leq h_n + r \leq H \quad \forall x \in B(\zeta_n, r), \quad \forall n \geq n_0. \tag{4.81}$$

Then, one can clearly find a smooth function v such that

$$v = \begin{cases} H & \text{on } B(\zeta, r + \delta), \\ 0 & \text{outside } B(\zeta, r + 2\delta), \end{cases} \tag{4.82}$$

this function v belongs then to the set defined by (4.76) (we see that $v \in K(\zeta, h)$ by
passing to the limit in (4.81)).

Step 2: Uniform bound for u_n. Using (4.15) for v given by (4.82), we obtain

$$\int_\Omega \nabla u_n \cdot \nabla (v - u_n)\, dx \geq 0 \quad \forall n \geq n_0$$

$$\Longrightarrow \quad ||\nabla u_n||_2^2 = \int_\Omega |\nabla u_n|^2\, dx \leq \int_\Omega \nabla u_n \cdot \nabla v\, dx \leq ||\nabla u_n||_2 ||\nabla v||_2 \tag{4.83}$$

where $|\cdot|_2$ denotes the $L_2(\Omega)$-norm, $|\cdot|$ the euclidean norm of the gradient. It follows that it holds that

$$||\nabla u_n||_2 \leq ||\nabla v||_2 \quad \forall n \geq n_0 \tag{4.84}$$

and thus u_n is bounded independently of n. Thus there exists a subsequence of n — still denoted by n such that for some $u_\infty \in H_0^1(\Omega)$ it holds that

$$u_n \rightharpoonup u_\infty \text{ in } H_0^1(\Omega), \quad u_n \to u_\infty \text{ in } L^2(\Omega), \quad u_n \to u_\infty \text{ a.e. in } \Omega, \tag{4.85}$$

(recall that $H_0^1(\Omega)$ is compactly imbedded in $L^2(\Omega)$).

Step 3: We have $u_\infty = u$. If we can show that $u_\infty = u$ then by uniqueness of this weak limit the whole sequence u_n will be converging toward u in $H_0^1(\Omega)$ weak. First, since $u_n \in K(\zeta_n, h_n)$, there exists a subset U_n of measure 0 such that

$$u_n(x) \geq \psi^{\zeta_n h_n}(x) \quad \text{a.e. } x \in B(\zeta_n, r) \setminus U_n. \tag{4.86}$$

Moreover there is a set V of measure 0 in Ω such that

$$u_n(x) \to u_\infty(x) \quad \text{in } \Omega \setminus V. \tag{4.87}$$

Set

$$U = (\cup_n U_n) \cup V. \tag{4.88}$$

Then U is a set of measure 0. Let $x \in B(\zeta, r) \setminus U$. For n large enough we have $x \in B(\zeta_n, r)\ \forall n$ and thus it holds that

$$u_n(x) \geq \psi^{\zeta_n h_n}(x) = h_n + \sqrt{r^2 - |x - \zeta_n|^2}. \tag{4.89}$$

Letting $n \to +\infty$ and using (4.87) we obtain

$$u_\infty(x) \geq \psi^{\zeta h}(x) \tag{4.90}$$

i.e. $u_\infty \in K(\zeta, h)$.

Next, we consider $u \in K(\zeta, h)$ the solution to (4.15). We set for n large enough

$$w_n = \begin{cases} u \vee \psi^{\zeta_n h_n} & \text{on } B\left(\zeta, r - \dfrac{\varepsilon}{2}\right), \\[2mm] u & \text{outside } B\left(\zeta, r - \dfrac{\varepsilon}{2}\right), \end{cases} \tag{4.91}$$

where ε is the real defined in (4.58) and $\vee$ denotes the maximum of two functions. We claim that for n large enough

$$w_n \in K(\zeta_n, h_n), \tag{4.92}$$

$$w_n \to u \text{ in } H_0^1(\Omega). \tag{4.93}$$

Indeed, for n large enough it holds that

$$B\left(\zeta, r - \frac{\varepsilon}{2}\right) \subset B\left(\zeta_n, r - \frac{\varepsilon}{3}\right) \subset B(\zeta_n, r) \tag{4.94}$$

so that the definition (4.91) makes sense. Moreover, since u and $\psi = \psi^{\zeta h}$ are continuous, for some $\delta > 0$ it holds that

$$u - \psi^{\zeta h} \geq \delta > 0 \text{ on } \Omega \setminus B\left(\zeta, r - \frac{2\varepsilon}{3}\right). \tag{4.95}$$

Now it is clear that on $B(\zeta, r - \frac{2\varepsilon}{3})$, $\psi^{\zeta_n h_n}$ converges uniformly toward $\psi^{\zeta h}$ and thus for n large enough it holds that

$$u - \psi^{\zeta_n h_n} \geq 0 \text{ on } \Omega \setminus B\left(\zeta, r - \frac{2\varepsilon}{3}\right). \tag{4.96}$$

Denote then by α a smooth function such that

$$\alpha = 1 \text{ on } B\left(\zeta, r - \frac{2\varepsilon}{3}\right), \qquad \alpha = 0 \text{ outside } B\left(\zeta, r - \frac{\varepsilon}{2}\right). \tag{4.97}$$

It is clear that it holds that

$$w_n = \alpha(u \vee \psi^{\zeta_n h_n}) + (1 - \alpha)u. \tag{4.98}$$

Recall that the maximum of two functions of H^1 belongs to H^1 (see [9]). Then from (4.94) we have that it holds that

$$u \vee \psi^{\zeta_n h_n} \in H^1\left(B\left(\zeta, r - \frac{\varepsilon}{2}\right)\right) \tag{4.99}$$

and thus $w_n \in H_0^1(\Omega)$. Moreover, combining (4.96) and the definition of w_n we have

$$w_n \in K(\zeta_n, h_n). \tag{4.100}$$

It is easy to see that – see also (4.94) –

$$\psi^{\zeta_n h_n} \to \psi^{\zeta h} \text{ in } H^1\left(B\left(\zeta, r - \frac{\varepsilon}{2}\right)\right) \tag{4.101}$$

and thus (see [9])

$$u \vee \psi^{\zeta_n h_n} \to u \text{ in } H^1\left(B\left(\zeta, r - \frac{\varepsilon}{2}\right)\right). \tag{4.102}$$

It follows from (4.98) that (4.93) holds. Thus from (4.15) we derive

$$\int_\Omega \nabla u_n \cdot \nabla (w_n - u_n)\, dx \geq 0$$

$$\Longrightarrow \quad \int_\Omega \nabla u_n \cdot \nabla w_n\, dx \geq \int_\Omega |\nabla u_n|^2\, dx. \tag{4.103}$$

Since $u_n \rightharpoonup u_\infty$ in $H_0^1(\Omega)$, taking the lim inf of both sides of the inequality we obtain

$$\int_\Omega \nabla u_\infty \cdot \nabla u\, dx \geq \int_\Omega |\nabla u_\infty|^2\, dx \tag{4.104}$$

(recall that the norm in a Hilbert space is lower semi continuous for the weak topology). That is to say we have

$$\int_\Omega \nabla u_\infty \cdot \nabla (u - u_\infty)\, dx \geq 0. \tag{4.105}$$

From (4.15) and (4.90) we also have

$$\int_\Omega \nabla u \cdot \nabla (u_\infty - u)\, dx \geq 0. \tag{4.106}$$

Thus, summing up (4.105), (4.106) it comes

$$\int_\Omega |\nabla (u - u_\infty)|^2\, dx \leq 0, \tag{4.107}$$

i.e., $u_\infty = u$.

Step 4: End of the proof. We have shown that

$$u_n \rightharpoonup u \quad \text{in} \quad H_0^1(\Omega). \tag{4.108}$$

To obtain the strong convergence, we go back to (4.103). Taking the lim sup we obtain

$$\limsup_{n \to +\infty} \int_\Omega |\nabla u_n|^2\, dx \leq \int_\Omega |\nabla u|^2\, dx.$$

Since – due to the lower semicontinuity of the norm for the weak topology

$$\liminf_{n \to +\infty} \int_\Omega |\nabla u_n|^2\, dx \geq \int_\Omega |\nabla u|^2\, dx$$

we have

$$\lim_{n \to +\infty} \int_\Omega |\nabla u_n|^2\, dx = \int_\Omega |\nabla u|^2\, dx$$

and the strong convergence of u_n towards u follows. This completes the proof of the lemma. $\qquad \square$

Then we can show:

Theorem 4.1. *The function $e = e(\zeta, h)$ defined by (4.16) is continuous on $\Omega_r \times \mathbb{R}$.*

Proof. Let $(\zeta_n, h_n) \in \Omega_r \times \mathbb{R}$ be a sequence of points such that (4.74) holds. Due to Lemma 4.4 we have

$$u_n = u(\zeta_n, h_n) \to u = u(\zeta, h) \quad \text{in } H_0^1(\Omega).$$

Thus, it holds

$$e(\zeta_n, h_n) = \frac{1}{2} \int_\Omega |\nabla u_n|^2 \, dx - Gh_n \to e(\zeta, h) = \frac{1}{2} \int_\Omega |\nabla u|^2 \, dx - Gh.$$

This completes the proof of the theorem.   □

In addition to the continuity of $e = e(\zeta, h)$ we can show:

Proposition 4.2. *Let e be the function defined by (4.16). Let v be a unit vector of $\mathbb{R}^2$. Then, for any $(\zeta, h) \in \Omega_r \times \mathbb{R}$ the function e is differentiable in the direction v and we have*

$$\frac{\partial e}{\partial v}(\zeta, h) = \lim_{\lambda \to 0} \frac{e(\zeta + \lambda v, h) - e(\zeta, h)}{\lambda} = -\int_\Lambda \frac{\partial \psi}{\partial v}(x) \Delta \psi(x) \, dx \qquad (4.109)$$

where Λ denotes the coincidence set defined by (4.30).

Proof. We give here a formal proof. Rigorous arguments can be found in [4]. Suppose that we can differentiate under the integral in (4.16), then we have:

$$\frac{\partial e}{\partial v}(\zeta, h) = \int_\Omega \nabla u \cdot \nabla \left(\frac{\partial u}{\partial v} \right) dx. \qquad (4.110)$$

For any $(\zeta, h) \in \Omega_r \times \mathbb{R}$ it holds that

$$u(\zeta, h)(x) = u(x) = 0 \quad \text{on } \partial \Omega.$$

Thus, for any v and λ small enough such that $(\zeta + \lambda h, h) \in \Omega_r \times \mathbb{R}$ we have

$$\frac{u(\zeta + \lambda v, h) - u(\zeta, h)}{\lambda} = 0 \quad \text{on } \partial \Omega.$$

Letting $\lambda \to 0$ we obtain

$$\frac{\partial u}{\partial v} = 0 \quad \text{on } \partial \Omega. \qquad (4.111)$$

Integrating by parts in (4.110) we get

$$\frac{\partial e}{\partial v}(\zeta, h) = -\int_\Omega \Delta u \frac{\partial u}{\partial v} \, dx$$

when of course all the above makes sense. Applying now (4.60) it comes

$$\frac{\partial e}{\partial v}(\zeta, h) = -\int_\Lambda \Delta \psi \frac{\partial \psi}{\partial v} \, dx$$

which completes the proof of the proposition. □

Proposition 4.2 will allow us to transform the minimization problem (4.17) to a minimization problem on a compact subset of $\Omega_r \times \mathbb{R}$. In addition we will need

Proposition 4.3. *It holds that*

$$\lim_{|h| \to +\infty} e(\zeta, h) = +\infty \tag{4.112}$$

uniformly in $\zeta \in \Omega_r$.

Proof. (a) When $h \leq -r$ it holds that

$$e(\zeta, h) = -Gh \tag{4.113}$$

and the result is clear.

(b) When $h \to +\infty$, from the Poincaré inequality we derive for some constant C independent of (ζ, h)

$$e(\zeta, h) \geq C \int_\Omega |u(x)|^2 \, dx - Gh \geq C \int_{B(\zeta, r)} |u(x)|^2 \, dx - Gh. \tag{4.114}$$

Since $u \in K(\zeta, h)$ we have for $h \geq 0$:

$$u(x) \geq \psi^{\zeta h}(x) \geq h \quad \text{on } B(\zeta, r)$$

and (4.114) becomes

$$e(\zeta, h) \geq C\pi r^2 h^2 - Gh. \tag{4.115}$$

Then letting $h \to +\infty$ the result follows. This completes the proof of the proposition. □

In order to reduce the minimization of e to a minimization on a compact subset of $\Omega_r \times \mathbb{R}$ we have to show that the ball is pushed away from the boundary of Ω when trying to minimize its energy. Considering a direction S passing through the center of the ball it is clear with the notation of Figure 25 that if the part of the membrane below S is smaller compared to the part above, the membrane will be stiff there and thus should push the ball away. This is what we would like to express mathematically in the lemma below.

Lemma 4.5. *Let $\zeta \in \Omega_r$ and v be a unit vector. Denote by R the reflection in the axis S going through ζ and orthogonal to v (see Figure 25). Assume that $h > -r$ and set*

$$\Omega_- = \{x \in \Omega \mid (x - \zeta) \cdot v < 0\},$$
$$\Omega_+ = \{x \in \Omega \mid (x - \zeta) \cdot v > 0\}, \tag{4.116}$$

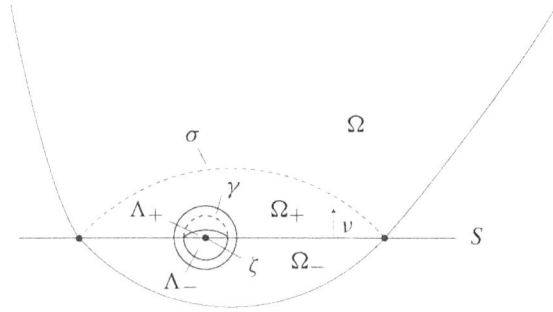

Figure 25. The membrane is pushing the ball inside

$$\Lambda_- = \Lambda \cap \Omega_-,$$
$$\Lambda_+ = \Lambda \cap \Omega_+. \tag{4.117}$$

(In (4.116) we denoted by a dot the usual scalar product in $\mathbb{R}^2$.) If

$$R(\Omega_-) \subsetneqq \Omega_+ \tag{4.118}$$

then it holds that

$$\Lambda_+ \subsetneqq R(\Lambda_-) \tag{4.119}$$

and moreover

$$\frac{\partial e}{\partial v}(\zeta, h) < 0. \tag{4.120}$$

Proof. On $R(\Omega_-)$, if u is the solution to (4.15), we define Ru by

$$Ru(x) = u(Rx). \tag{4.121}$$

Consider now the open subset

$$E = R(\Omega_-) \setminus R(\Lambda_-). \tag{4.122}$$

Its boundary is divided in three parts that we denote by S, γ, σ – see Figure 25. We know – see (4.60) – that it holds that

$$-\Delta u = 0 \quad \text{in } \Omega_- \setminus \Lambda_- \tag{4.123}$$

and

$$-\Delta u \geq 0 \quad \text{in } \Omega. \tag{4.124}$$

Since $h > -r$, we cannot have $u \equiv 0$ and thus since $u \geq 0$ (see (4.25)), by the strong maximum principle, it holds that

$$u > 0 \quad \text{in } \Omega. \tag{4.125}$$

Since the Laplace operator is invariant by reflection we have

$$-\Delta(Ru - u) = -\Delta Ru + \Delta u = -\Delta u(Rx) + \Delta u(x) = \Delta u(x) \leq 0 \quad \text{in } E, \quad (4.126)$$

by (4.122)–(4.124). Moreover on the boundary of E it holds that

$$(Ru - u)(x) = u(Rx) - u(x) = \begin{cases} 0 - u(x) < 0 & \text{on } \sigma, \\ \psi(x) - u(x) \leq 0 & \text{on } \gamma, \\ u(x) - u(x) = 0 & \text{on } S, \end{cases} \quad (4.127)$$

(recall that $\psi(Rx) = \psi(x)$). Thus, by the strong maximum principle, it holds that

$$Ru - u < 0 \quad \text{in } E. \quad (4.128)$$

Let us show now that (4.119) holds. For that let us consider

$$x \in \Lambda_+.$$

We have

$$u(x) = \psi(x) \quad \Longrightarrow \quad u(Rx) \geq \psi(Rx) = \psi(x) = u(x) \quad \Longrightarrow \quad Ru - u(x) \geq 0.$$

By (4.128) this implies that $x \notin E$. Thus $x \in R(\Lambda_-)$ and we have

$$\Lambda_+ \subset R(\Lambda_-). \quad (4.129)$$

If we had

$$\Lambda_+ = R(\Lambda_-)$$

then we would have

$$Ru - u = 0 \quad \text{on } \gamma$$

and also

$$\nabla Ru = \nabla R\psi = \nabla \psi = \nabla u \quad \text{on } \Gamma$$

(recall that u is C^1). But this contradicts the Hopf maximum principle, and thus (4.119) holds. We now turn to (4.120). Using (4.109) we have

$$\frac{\partial e}{\partial \nu}(\zeta, h) = -\int_{\Lambda_+} \frac{\partial \psi}{\partial \nu} \Delta \psi \, dx - \int_{\Lambda_-} \frac{\partial \psi}{\partial \nu} \Delta \psi \, dx$$

$$= -\int_{\Lambda_+} \frac{\partial \psi}{\partial \nu} \Delta \psi \, dx - \int_{R(\Lambda_+)} \frac{\partial \psi}{\partial \nu} \Delta \psi \, dx - \int_{\Lambda_- \setminus R(\Lambda_+)} \frac{\partial \psi}{\partial \nu} \Delta \psi \, dx$$

$$(4.130)$$

((4.129) implies $R(\Lambda_+) \subset \Lambda_-$).
By (4.2) we have

$$\frac{\partial \psi}{\partial \nu}(x) = \nu \cdot \nabla_\zeta \psi(x) = \frac{\nu \cdot (x - \zeta)}{\sqrt{r^2 - |x - \zeta|^2}} \quad (4.131)$$

(recall that we are differentiating in ζ). In particular

$$\frac{\partial \psi}{\partial v}(Rx) = \frac{v \cdot (Rx - \zeta)}{\sqrt{r^2 - |x - \zeta|^2}} = \frac{v \cdot R(x - \zeta)}{\sqrt{r^2 - |x - \zeta|^2}} = -\frac{\partial \psi}{\partial v}(x). \tag{4.132}$$

Thus making the change of variable $x \to Rx$ in the second integral we have

$$\int_{R(\Lambda_+)} \frac{\partial \psi}{\partial v}(x)\Delta\psi(x)\,dx = \int_{\Lambda_+} \frac{\partial \psi}{\partial v}(Rx)\Delta\psi(Rx)\,dx = -\int_{\Lambda_+} \frac{\partial \psi}{\partial v}(x)\Delta\psi(x)\,dx$$

(recall that $\Delta\psi(Rx) = \Delta\psi(x)$ – see (4.51)). Thus from (4.130) we derive

$$\frac{\partial e}{\partial v}(\zeta, h) = -\int_{\Lambda_- \setminus R(\Lambda_+)} \frac{\partial \psi}{\partial v} \cdot \Delta\psi\,dx. \tag{4.133}$$

By (4.131), (4.51) we have

$$\frac{\partial \psi}{\partial v}(x) < 0, \quad \Delta\psi < 0 \text{ in } \Lambda_-$$

and (4.120) follows from (4.133). This completes the proof of the lemma.   □

We can now prove the following existence result.

Theorem 4.2. *Let us assume that we can find ε such that*

$$\forall r' \in (r, r + \varepsilon), \forall \zeta \text{ such that } \mathrm{dist}(\zeta, \Gamma) = r' \text{ there}$$
$$\text{exists } v \text{ such that the situation of Lemma 4.5 holds.} \tag{4.134}$$

Then there exists a point minimizing e on $\Omega_r \times \mathbb{R}$ and thus a solution to the problems (4.11), (4.17).

Proof. For $h \leq -r$ we have $e(\zeta, h) = -Gh$ and thus

$$\frac{\partial e}{\partial v}(\zeta, h) = 0.$$

By Proposition 4.3 to minimize e on $\Omega_r \times \mathbb{R}$ it is enough to minimize e on

$$\Omega_r \times [-r, h^*]$$

for some positive h^*. Now due to (4.134) we have for $h > -r$

$$\frac{\partial e}{\partial v}(\zeta, h) < 0$$

for any $\zeta \in \Omega_r$ such that $\mathrm{dist}(\zeta, \Gamma) < r + \varepsilon$. Clearly at such a point e cannot achieve a minimum. Thus, to minimize e on $\Omega_r \times \mathbb{R}$, we are reduced to minimize e on

$$\{\zeta \in \Omega \mid \mathrm{dist}(\zeta, \Gamma) \geq r + \varepsilon\} \times [-r, h^*]$$

i.e. on a compact set and the existence of a point realizing the minimum is due to the continuity of e established in Theorem 4.1. This completes the proof of the theorem.   □

Remark 4.2. There are many situations where (4.134) holds. For instance this is the case when Ω is convex, smooth and r small enough. Some non-convex situations could also be handled for instance it is clear that (4.134) holds for r small enough in the case of Figure 26.

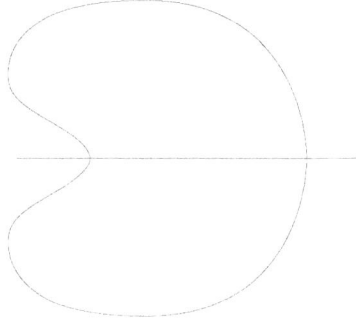

Figure 26. A non-convex case where existence holds

Remark 4.3. The property (4.120) could be used to locate the position of the equilibrium. Indeed consider for instance the convex domain described in the figure below. Then for any v orthogonal to the axis of symmetry and as in Figure 27 we have:

$$\frac{\partial e}{\partial v}(\zeta, h) < 0.$$

Thus the minimum of the energy is necessarily achieved on Σ the axis of symmetry

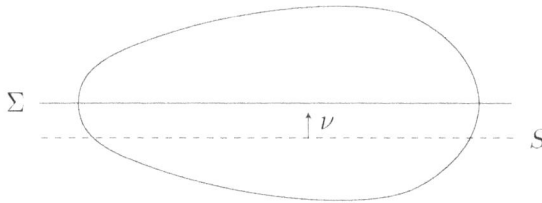

Figure 27. A convex case

of Ω. In the case of an ellipse or a disk this shows that the minimum of the energy is necessarily achieved at the center of the ellipse or the disk.

In the case of a circular membrane we can show

Theorem 4.3. *Suppose that* $\Omega = B(0, R)$ *the open ball of center 0 and radius R. Set* $g = G/2\pi$. *Then the minimum of e is achieved at the single point*

$$\zeta = 0, \quad h = -g \ln \frac{s}{R} - \sqrt{r^2 - s^2} \tag{4.135}$$

where s is given by

$$s^2 = \{-g^2 + \sqrt{4g^2r^2 + g^4}\}/2. \tag{4.136}$$

Moreover $u = u(\zeta, h)$ is given by (4.35).

Proof. It follows from Remark 4.3 that $\zeta = 0$. Next we can compute for $h > -r$, $\frac{\partial e}{\partial h}(\zeta, h)$ proceeding as we did for establishing (4.109). Indeed, differentiating formally under the integral symbol we have

$$\frac{\partial e}{\partial h}(\zeta, h) = \int_\Omega \nabla u \cdot \nabla \frac{\partial u}{\partial h} dx - G$$

(this can be justified, see [4]). Since – compare to (4.111)

$$\frac{\partial u}{\partial h} = 0 \quad \text{on } \Gamma,$$

integrating by parts we get

$$\frac{\partial e}{\partial h}(\zeta, h) = -\int_\Omega \Delta u \frac{\partial u}{\partial h} dx - G = -\int_\Lambda \Delta \psi \frac{\partial \psi}{\partial h} dx - G$$
$$= \int_\Lambda \Delta \psi \, dx - G = -\int_\Omega \Delta u \, dx - G = -\int_\Gamma \frac{\partial u}{\partial n} d\sigma(x) - G \tag{4.137}$$

where $\frac{\partial u}{\partial n}$ denotes the outside normal derivative to Γ, $d\sigma(x)$ the superficial measure on Γ. Now, clearly for any h the solution $u = u(0, h)$ is given by (4.34), (4.35) where we have set $\zeta = 0$. To determine h we remark that the minimum of e can be achieved only when (4.137) vanishes – that is to say when

$$-2\pi R u'(R) = G \quad \Longleftrightarrow \quad -R u'(R) = g. \tag{4.138}$$

Using (4.35) we derive that

$$s^2 = g\sqrt{r^2 - s^2} \quad \Longleftrightarrow \quad s^4 + g^2 s^2 - g^2 r^2 = 0 \tag{4.139}$$

which clearly implies (4.136). To get (4.135) we remark that by (4.138)

$$u = -g \ln \frac{\varrho}{R} \quad \text{if } \varrho \geq s. \tag{4.140}$$

For $\varrho = s$ we obtain – since $u = \psi^{\zeta h}$:

$$-g \ln \frac{s}{R} = h + \sqrt{r^2 - s^2}$$

which completes the proof of the theorem.   □

Remark 4.4. Note that s is independent of R and that it holds that

$$\lim_{R \to +\infty} h = +\infty. \tag{4.141}$$

Remark 4.5. One can compute the value of the energy at the minimum. Indeed we have

$$e(\zeta, h) = \frac{1}{2} \int_{\Omega} |\nabla u|^2 \, dx - Gh = \frac{1}{2} \int_{\Omega} -\Delta u u \, dx - Gh = -\frac{1}{2} \int_{\Lambda} \Delta \psi \psi \, dx - Gh.$$

Since ψ is given by (4.2) it comes using polar coordinates

$$e(\zeta, h) = -\frac{1}{2} \int_{\Lambda} \Delta \psi \{h + \sqrt{r^2 - \varrho^2}\} \, dx - Gh$$

$$= -\frac{1}{2} \int_{\Omega} \Delta u \cdot h - Gh - \frac{1}{2} \int_{\Lambda} \frac{\varrho^2 - 2r^2}{r^2 - \varrho^2} \varrho \, d\varrho \, d\theta$$

(recall (4.51)). Using the Green formula again we derive – see (4.137)

$$e(\zeta, h) = -\frac{h}{2} \int_{\Gamma} \frac{\partial u}{\partial n} \, d\sigma(x) - Gh - \pi \int_{0}^{s} \frac{\varrho^2 - 2r^2}{r^2 - \varrho^2} \varrho \, d\varrho$$

$$= -\frac{Gh}{2} - \pi \int_{0}^{s} \frac{\varrho^2 - 2r^2}{r^2 - \varrho^2} \cdot \varrho \, d\varrho$$

$$= -\frac{Gh}{2} - \frac{\pi}{2} \int_{0}^{s^2} \frac{\xi - 2r^2}{r^2 - \xi} \, d\xi \qquad \qquad (4.142)$$

$$= -\frac{Gh}{2} - \frac{\pi}{2} \int_{0}^{s^2} \left\{-1 - \frac{r^2}{r^2 - \xi}\right\} d\xi$$

$$= -\frac{Gh}{2} - \frac{\pi}{2} \{-\xi + r^2 \ln(r^2 - \xi)\}|_{0}^{s^2}$$

$$= -\frac{Gh}{2} + \frac{\pi}{2} \{s^2 - r^2 \ln(1 - s^2/r^2)\}.$$

It is clear from this formula and Remark 4.4 that

$$\lim_{R \to +\infty} e(\zeta, h) = -\infty. \qquad \qquad (4.143)$$

4.3 Uniqueness issue

In this section we would like to show that the minimum of e in $\Omega_r \times \mathbb{R}$ is not necessarily unique. For that we consider the domain

$$\Omega = \{(-S, S) \times (-1, 1)\} \cup B(-S, R) \cup B(S, R) \qquad (4.144)$$

where S is given by

$$S = R + 1 \qquad \qquad (4.145)$$

(see Figure 28). Then we can show

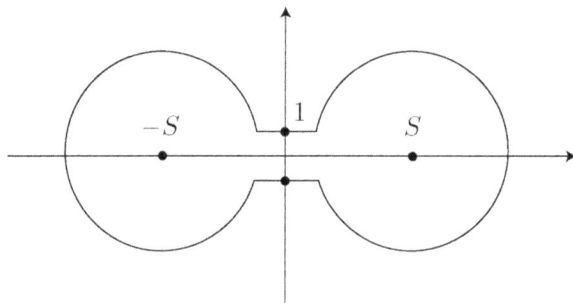

Figure 28. A domain where uniqueness fails

Theorem 4.4. *Let Ω be the domain defined by (4.144). Then, for*

$$r < 1, \quad R \text{ large enough,} \tag{4.146}$$

the minimization problem admits at least two solutions.

Proof. Suppose that (4.17) admits a unique solution. By Remark 4.3 the minimum of the energy is achieved on the axis $\zeta_2 = 0$. For symmetry reasons we have

$$e(-\zeta_1, 0, h) = e(\zeta_1, 0, h)$$

and if the minimum is unique we have necessarily

$$\zeta_1 = 0.$$

Let us denote by $(u_R, 0, h_R)$ the unique solution to (4.11). If Ω is the square

$$Q = (-1, 1)^2$$

then there exists a solution $(u_Q, 0, h_Q)$ to the corresponding problem (4.11). If we extend u_Q by 0 outside Q, then u_Q is an admissible deformation for (4.11) associated to Ω given by (4.144). Thus we have

$$E(u_Q; 0, h_Q) = \frac{1}{2} \int_Q |\nabla u_Q|^2 \, dx - G h_Q \geq E(u_R; 0, h_R)$$

$$= \frac{1}{2} \int_\Omega |\nabla u_R|^2 \, dx - G h_R \geq \frac{1}{2} \int_Q |\nabla u_R|^2 \, dx - G h_R. \tag{4.147}$$

Since $u_R = 0$ on $(-1, 1) \times \{-1, 1\}$, by the Poincaré inequality we have for some constant c independent of R

$$E(u_Q; 0, h_Q) \geq \frac{c}{2} \int_Q u_R^2 \, dx - G h_R$$

$$\geq \frac{c}{2} \int_{B(0,r)} h_R^2 \, dx - G h_R = \frac{\pi}{2} cr^2 h_R^2 - G h_R. \tag{4.148}$$

if $h_R \geq 0$ (we know by the existence part that $h_R \geq -r$). Thus from (4.148) we deduce that h_R is bounded independently of R. But then it follows that

$$E(u_R; 0, h_R) \geq -Gh_R \geq -M$$

independently of R. If now we consider the deformation u_B corresponding to the minimization problem (4.11) for $\Omega = B(S, R+1)$ and if we extend this deformation by 0 outside of $B = B(S, R+1)$, u_B becomes an admissible deformation for the problem corresponding to Ω defined by (4.144) and we have

$$\frac{1}{2} \int_B |\nabla u_B|^2 \, dx - Gh_B \geq -M$$

(h_B denotes the altitude of the equilibrium for $\Omega = B$). But since M is independent of R, by (4.143), we get a contradiction. Thus the equilibrium position of the ball cannot be unique for R large. This completes the proof of the theorem. □

4.4 Some open questions

It would be of course very interesting to remove the assumption (4.134) in the existence result. One could consider also a nonhomogeneous ball – compare to (2.11).

Regarding uniqueness some interesting issues have yet to be addressed. In the case where Ω is convex, we do not know if e admits at most one minimizer. For example the problem of uniqueness in the case of Figure 27 is open. Some interesting issues of non uniqueness remain also. We do not know if nonuniqueness holds in the case of Figure 26 even so it is a reasonable guess. More generally it would be interesting to find out conditions insuring non uniqueness of minimizers. Of course the above issues have to be addressed also in the case of the energy defined by (4.18) or in the case of nonhomogeneous energy of the type

$$\frac{1}{2} \int_\Omega \sigma(x) |\nabla u(x)|^2 \, dx - Gh,$$

– see (2.10). Suppose now that

$$\Omega = (-\ell, \ell) \times (-1, 1)$$

where $\ell > 1$ and that $r < 1$. Then, the problem of minimization (4.11) admits certainly a unique solution (for symmetry reason we must have $\zeta = 0$, see Figure 29). Then, it would be interesting to examine what happens when $\ell \to \infty$. For instance does the coincidence set shrink to a curve? What is at the limit the relationship between this problem and the problem of a disk on a wire? The undeformed configuration of the wire being here $(-1, 1)$ – cf. [10] for all these types of questions.

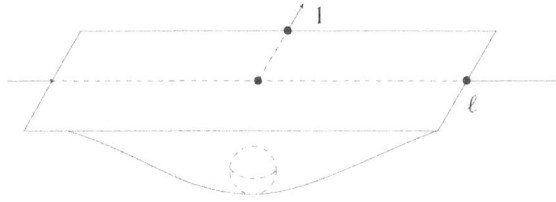

Figure 29. The rectangle case with one dimension large

5 The case of several balls on an elastic membrane

For simplicity we restrict ourselves to the case of two balls. We denote by r_1, r_2 the radii of the balls and by G_1, G_2 their weights. The position of the balls will be

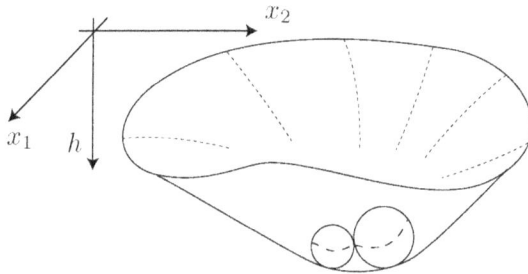

Figure 30. The equilibrium position of two balls rolling on a membrane

completely determined by the position of their centers

$$C_1 = (\zeta_1, h_1), \quad C_2 = (\zeta_2, h_2), \tag{5.1}$$

$\zeta_i \in \mathbb{R}^2$, $h_i \in \mathbb{R}$, $i = 1, 2$. If Ω denotes the undeformed configuration of the elastic membrane at hand we will suppose that the balls cannot touch the rim of the membrane i.e. that

$$\zeta_i \in \Omega_{r_i} = \{ x \in \Omega \mid \text{dist}(x, \Gamma) > r_i \}, \quad i = 1, 2. \tag{5.2}$$

Moreover we will assume that these balls cannot penetrate each other in such a way that

$$|\zeta_1 - \zeta_2|^2 + (h_1 - h_2)^2 \geq (r_1 + r_2)^2, \tag{5.3}$$

$|\cdot|$ denotes the euclidean norm in $\mathbb{R}^2$.

The bottom parts of the balls will be described by the functions

$$\psi_i = \psi^{\zeta_i h_i}(x) = h_i + \sqrt{r_i^2 - |x - \zeta_i|^2}, \quad x \in B(\zeta_i, r_i), \ i = 1, 2 \tag{5.4}$$

(see (4.3), (4.4) for the notation). Then, for a position

$$\xi = (\zeta_1, h_1, \zeta_2, h_2) \in \mathbb{R}^6 \tag{5.5}$$

of the centers an admissible deformation is an element $u \in K = K(\xi)$ defined by

$$K(\xi) = \{ v \in H_0^1(\Omega) \mid v(x) \geq \psi^{\zeta_i, h_i}(x) \text{ a.e. } x \in B(\zeta_i, r_i), \ i = 1, 2 \}. \tag{5.6}$$

Thus for

$$\xi = (\zeta_1, h_1, \zeta_2, h_2) \in \Omega_{r_1} \times \mathbb{R} \times \Omega_{r_2} \times \mathbb{R} \tag{5.7}$$

and u an admissible deformation the energy of the system is given by

$$E(u; \xi) = \frac{1}{2} \int_\Omega |\nabla u(x)|^2 \, dx - G_1 h_1 - G_2 h_2. \tag{5.8}$$

The minimization problem that we would like to address is then to find u^0, ξ^0 such that it holds that

$$\begin{cases} \xi^0 = (\zeta_1^0, h_1^0, \zeta_2^0, h_2^0) \in \Omega_{r_1} \times \mathbb{R} \times \Omega_{r_2} \times \mathbb{R}, \\ \xi^0 \text{ satisfies } (5.3), \ u^0 \in K(\xi^0), \\ E(u^0; \xi^0) \leq E(u; \xi) \ \forall \xi \in \Omega_{r_1} \times \mathbb{R} \times \Omega_{r_2} \times \mathbb{R}, \\ \xi \text{ satisfying } (5.3), \ \forall u \in K(\xi). \end{cases} \tag{5.9}$$

Arguing as in the preceding sections, we introduce $u = u(\xi)$ the solution to the variational inequality

$$\begin{cases} u \in K(\xi), \\ \int_\Omega \nabla u \cdot \nabla(v - u) \, dx \geq 0 \quad \forall v \in K(\xi). \end{cases} \tag{5.10}$$

Then for $\xi \in \Omega_{r_1} \times \mathbb{R} \times \Omega_{r_2} \times \mathbb{R}$, ξ satisfying (5.3), $u = u(\xi)$ we set

$$e(\xi) = \frac{1}{2} \int_\Omega |\nabla u(x)|^2 \, dx - G_1 h_1 - G_2 h_2. \tag{5.11}$$

Then, clearly, the problem (5.9) is equivalent to minimize e on a subset of $\Omega_{r_1} \times \mathbb{R} \times \Omega_{r_2} \times \mathbb{R}$, i.e. to find ξ^0 such that

$$\begin{cases} \xi^0 \in \Omega_{r_1} \times \mathbb{R} \times \Omega_{r_2} \times \mathbb{R}, & \xi^0 \text{ satisfies } (5.3), \\ e(\xi^0) \leq e(\xi) \ \forall \xi \in \Omega_{r_1} \times \mathbb{R} \times \Omega_{r_2} \times \mathbb{R}, & \xi \text{ satisfying } (5.3). \end{cases} \tag{5.12}$$

This problem is in general open. Let us collect here the results that are available with the techniques we have developed above.

First let us denote by

$$u_i = u(\zeta_i, h_i) \tag{5.13}$$

the solution to (4.15). Then we have

Proposition 5.1. *Let* $\xi = (\zeta_1, h_1, \zeta_2, h_2) \in \Omega_{r_1} \times \mathbb{R} \times \Omega_{r_2} \times \mathbb{R}$ *and* $u = u(\xi)$ *be the solution to* (5.10). *Then it holds that*

$$u(\xi) \geq u_i = u(\zeta_i, h_i) \quad in \ \Omega, \ i = 1, 2. \tag{5.14}$$

Proof. It is enough to show that

$$(u_i - u)^+ = 0, \quad i = 1, 2. \tag{5.15}$$

For that remark that

$$u + (u_i - u)^+ \in K(\xi). \tag{5.16}$$

Thus from (5.10) we deduce

$$\int_\Omega \nabla u \cdot \nabla (u_i - u)^+ \, dx \geq 0. \tag{5.17}$$

Next since

$$u_i - (u_i - u)^+ = u \ \text{ or } \ u_i$$

we have clearly

$$u_i - (u_i - u)^+ \in K(\zeta_i, h_i)$$

and from (4.15) we obtain

$$-\int_\Omega \nabla u_i \cdot \nabla (u_i - u)^+ \, dx \geq 0. \tag{5.18}$$

Adding (5.17) and (5.18) we get

$$\int_\Omega |\nabla (u_i - u)^+| \, dx = \int_\Omega \nabla (u_i - u) \cdot \nabla (u_i - u)^+ \leq 0$$

and (5.15) follows. This completes the proof of the proposition. $\qquad\square$

As a consequence we can show

Proposition 5.2. *Let* $\xi = (\zeta_1, h_1, \zeta_2, h_2) \in \Omega_{r_1} \times \mathbb{R} \times \Omega_{r_2} \times \mathbb{R}$. *Let* $u = u(\xi)$ *the solution to* (5.10). *Then it holds that*

$$u \in W^{1,\infty}(\Omega). \tag{5.19}$$

The coincidence set

$$\Lambda = \{ x \in B(\zeta_1, r_1) \cup B(\zeta_2, r_2) \mid u(x) = \psi_1(x) \ or \ u(x) = \psi_2(x) \} \tag{5.20}$$

is a closed subset of Ω *such that for some* $\varepsilon > 0$ *it holds that*

$$\Lambda \subset B(\zeta_1, r_1 - \varepsilon) \cup B(\zeta_2, r_2 - \varepsilon). \tag{5.21}$$

Moreover we have

$$\Delta u \geq 0 \quad in \ \Omega, \tag{5.22}$$

$$\Delta u = 0 \quad in \ \Omega \setminus \Lambda. \tag{5.23}$$

Proof. This is an easy consequence of Proposition 4.1 and 5.1. $\qquad \square$

In arguing as in Theorem 3.1 we can prove

Theorem 5.1. *The function e is continuous at any point of $\Omega_{r_1} \times \mathbb{R} \times \Omega_{r_2} \times \mathbb{R}$ satisfying (5.3).*

At this stage many open questions remain. We refer the reader to [16]. Indeed the system formed by two balls has not the symmetry of a single ball. For instance, if we can find an analogue of the differentiation formula (4.109), to use it as in Lemma 4.5 is a different story. So, it is not so easy to reduce the minimization problem (5.12) to a minimization on a compact subset of $\mathbb{R}^6$ – even in the case where Ω is a circular membrane and the balls identical. Considering – see Lemma 4.5 – an axis S going through the center of the balls one can show that one decreases the energy when S moves toward the center of the membrane (see [16], [12]). However, we are not yet able to conclude that (5.12) achieves a minimum. The nature of this equilibrium is more mysterious. For instance do we have for small weights the two balls sitting at the same level in the middle of the membrane? Do they tilt when the weight increases (see Theorem 3.2)? Could we have the situation of Figure 17 when the balls are not identical and the small ball becomes heavy? Could we show then that the equilibrium is achieved when the two balls are centered in the middle of the membrane and that it is unique? The issue of uniqueness of a solution is, in the case of two balls, more subtle also than in the case of a single ball. For instance in the case of a circular membrane uniqueness should hold up to a rotation centered at the center of the membrane. Thus even for identical balls and simple domain Ω the problem of existence, uniqueness and nature of the equilibrium is far from being trivial. To convince the reader further let us consider the domain Ω_ℓ introduced in the preceding section and given by

$$\Omega_\ell = (-\ell, \ell) \times (-1, 1), \quad \ell \geq 1. \tag{5.24}$$

Let us drop two identical balls on this membrane. What is the equilibrium position(s)? What is going to happen for $\ell = 1$ i.e. for a square membrane? What will follow if we let $\ell > 1$? In what direction will the balls be lined up? What will happen when $\ell \to +\infty$. The questions are very simple. However, to be tackled they need some skills. We hope the reader will be eager to attack them and develop new elegant techniques that can be used in the wide topic of elliptic problems and calculus of variations.

Acknowledgements: This work has been supported by the Swiss National Science Foundation under the contract 20-67618.02. This help is greatly acknowledged.

References

[1]   A. Aissani, Thesis, University of Metz, 2000.

[2]   A. Aissani, M. Chipot, On the equilibrium position of a square on an elastic wire, *J. Convex Anal.* 8 (2001), 171–195.

[3]   A. Aissani, M. Chipot, S. Fouad, On the deformation of an elastic wire by one or two heavy disks, *Arch. Math.* 76 (2001), 467–480.

[4]   J. Bemelmans, M. Chipot, On a variational problem for an elastic membrane supporting a heavy ball, *Calc. Var. Partial Differential Equations* 3 (1995), 447–473.

[5]   H. Brezis, Problèmes unilatéraux, *J. Math. Pures Appl.* 51 (1972), 1–168.

[6]   H. Brezis, *Analyse fonctionnelle*, Masson, Paris 1983.

[7]   H. Brezis, D. Kinderlehrer: The smoothness of solutions to nonlinear variational inequalities, *Indiana Univ. Math. J.* 23 (1974), 831–844.

[8]   M. Chipot, *Variational inequalities and flows in porous media*, Appl. Math. Sci. 52, Springer-Verlag, New York 1984.

[9]   M. Chipot, *Elements of Nonlinear Analysis*, Birkhäuser, 2000.

[10] M. Chipot, ℓ *goes to plus infinity*, Birkhäuser, 2001.

[11] M. Chipot, On the equilibrium positions of two heavy disks supported by an elastic wire, to appear.

[12] M. Chipot, in preparation.

[13] R. Dautray, J. L. Lions, *Mathematical analysis and numerical methods for science and technology*, Springer-Verlag, 1982.

[14] C. M. Elliott, A. Friedman, The contact set of a rigid body partially supported by a membrane, *Nonlinear Anal.* 10 (1986), 251–276.

[15] A. Friedman, *Variational Principles and Free-Boundary Problems*, Wiley, New York 1982.

[16] S. Fouad, Thesis, University of Metz, 1999.

[17] D. Gilbarg, N. S. Trudinger, *Elliptic Partial Differential Equations of Second Order*, Grundlehren Math. Wiss. 224, Springer-Verlag, Berlin, Heidelberg, New York 1977.

[18] D. Kinderlehrer, G. Stampacchia, *An introduction to variational inequalities and their applications*, Acadademic Press, 1980.

[19] J. F. Rodrigues, *Obstacle problems in mathematical physics*, North-Holland Math. Stud. 134, North Holland, Amsterdam 1987.

Scattering, spreading, and localization of an acoustic pulse by a random medium

Josselin Garnier

Laboratoire de Statistique et Probabilités
Université Paul Sabatier
118 Route de Narbonne, 31062 Toulouse Cedex 4, France
email: garnier@cict.fr

Abstract. Random media have material properties with such complicated spatial variations that they can only be described statistically. When looking at waves propagating in these media, we can only expect in general a statistical description of the wave. But sometimes there exists a deterministic result: the wave dynamics only depends on the statistics of the medium, and not on the particular realization of the medium. Such a phenomenon arises when the different scales present in the problem (wavelength, correlation length, and propagation distance) can be separated. In this lecture we restrict ourselves to one-dimensional wave problems that arise naturally in acoustics and geophysics.

1 Introduction

Wave propagation in linear random media has been studied for a long time by perturbation techniques when the random inhomogeneities are small. One finds that the mean amplitude decreases with distance traveled, since wave energy is converted to incoherent fluctuations. The fluctuating part of the field intensity is calculated approximatively from a transport equation, a linear radiative transport equation. This theory is well-known [21], although a complete mathematical theory is still lacking (for the most recent developments, see for instance [14]). However this theory is known to be false in one-dimensional random media. This was first noted by Anderson [1], who claimed that random inhomogeneities trap wave energy in a finite region and do not allow it to spread as it would normally. This is the so-called wave localization phenomenon. It was first proved mathematically in [20]. Extensions and generalizations follow these pioneer works so that the problem is now well understood [10]. The mathematical statement is that the spectrum of the reduced wave equation is pure point with exponentially decaying eigenfunctions. However the authors did not give quantitative information associated with the wave propagation as no exact solution is

available. In this lecture we are not interested in the study of the strongest form of
Anderson localization. We actually address the simplest form of this problem: the
wave transmission through a slab of random medium. It is now well-known that the
transmission of the slab tends exponentially to zero as the length of the slab tends
to infinity. Furstenberg first treated discrete versions of the transmission problem
[18], and finally Kotani gave a proof of this result with minimal hypotheses [25].
The connection between the exponential decay of the transmission and the Anderson
localization phenomenon is clarified in [13]. Once again, these works deal with qual-
itative properties. Quantitative information can be obtained only for some asymptotic
limits: large or small wavenumbers, large or small variances of the fluctuations of the
parameters of the medium, etc. A lot of work was devoted to the quantitative analysis
of the transmission problem, in particular by Rytov, Tatarski, Klyatskin [23], and by
Papanicolaou and its co-workers [24]. The tools for the quantitative analysis are limit
theorems for stochastic equations developed by Khasminskii [22] and by Kushner
[27].

There are three basic length scales in wave propagation phenomena: the typical
wavelength λ, the typical propagation distance L, and the typical size of the inhomogeneities l_c. There is also a typical order of magnitude that characterizes the standard
deviation of the dimensionless fluctuations of the parameters of the medium. It is not
always so easy to identify the scale l_c, but we may think of l_c as a typical correlation
length. When the standard deviation of the relative fluctuations is small $\varepsilon \ll 1$, then
the most effective interaction of the waves with the random medium will occur when
$l_c \sim \lambda$, that is, the wavelength is comparable to the correlation length. Such an interaction will be observable when the propagation distance L is large ($L \sim \lambda \varepsilon^{-2}$). This
is the typical configuration in optics and in optical fibers.

Throughout this lecture we shall consider scalings arising in acoustics and geophysics. The main differences with optics are that the fluctuations are not small, but
the typical wavelength of the pulse $\lambda \sim 150$ m is small compared to the probing
depth $L \sim 10 - 50$ km, but large compared to the correlation length $l_c \sim 2 - 3$ m
[38]. Accordingly we shall introduce a small parameter $0 < \varepsilon \ll 1$ and consider
$l_c \sim \varepsilon^2$, $\lambda \sim \varepsilon$, and $L \sim 1$. The parameter ε is the ratio of the typical wavelength
to propagation depth, as well as the ratio of correlation length to wavelength. This is
a particularly interesting scaling limit mathematically because it is a high frequency
limit with respect to the large scale variations of the medium, but it is a low frequency
limit with respect to the fluctuations, whose effect acquires a canonical form independent of details. We shall study the asymptotic behavior of the scattered wave in the
framework introduced by Papanicolaou based on the separation of these scales.

The lecture is organized as follows. In Section 2 we present the method of averaging
for stochastic processes that is an extension of the law of large numbers for the sums
of independent random variables. These results provide the tools for the effective
medium theory developed in Section 3. We give a review of the properties of Markov
processes in Section 4. We focus our attention to ergodic properties, and we propose
limit theorems for ordinary differential equations driven by Markov processes that

are applied in the following sections. Section 5 is devoted to the O'Doherty–Anstey problem, that is to say the spreading of a pulse traveling through a random medium. We compute the localization length of a monochromatic pulse in Section 6. We finally study the exponential localization phenomenon for a pulse in Section 7 and show that it is a self-averaging process.

2 Averages of stochastic processes

We begin by a brief review of the two main limit theorems for sums of independent random variables.

• The Law of Large Numbers: If $(X_i)_{i \in \mathbb{N}}$ is a sequence of independent identically distributed $\mathbb{R}$-valued random variables, with $\mathbb{E}[|X_1|] < \infty$, then the normalized partial sums

$$S_n = \frac{1}{n} \sum_{k=1}^{n} X_k$$

converge to the statistical average $\bar{X} = \mathbb{E}[X_1]$ with probability one (write a.s. for almost surely).

• The Central Limit Theorem: If $(X_i)_{i \in \mathbb{N}}$ is a sequence of independent and identically distributed $\mathbb{R}$-valued random variables, with $\bar{X} = \mathbb{E}[X_1]$ and $\mathbb{E}[X_1^2] < \infty$, then the normalized partial sums

$$\tilde{S}_n = \frac{1}{\sqrt{n}} \sum_{k=1}^{n} (X_k - \bar{X})$$

converge in distribution to a Gaussian random variable with mean 0 and variance $\sigma^2 = \mathbb{E}[(X_1 - \bar{X})^2]$. This distribution is denoted by $\mathcal{N}(0, \sigma^2)$. The above assertion means that, for any continuous bounded function f,

$$\mathbb{E}[f(\tilde{S}_n)] \xrightarrow{n \to \infty} \frac{1}{\sqrt{2\pi\sigma^2}} \int_{-\infty}^{\infty} f(x) \exp\left(-\frac{x^2}{2\sigma^2}\right) dx$$

or, for any interval $I \subset \mathbb{R}$,

$$\mathbb{P}(\tilde{S}_n \in I) \xrightarrow{n \to \infty} \frac{1}{\sqrt{2\pi\sigma^2}} \int_{I} \exp\left(-\frac{x^2}{2\sigma^2}\right) dx.$$

2.1 A toy model

Let us consider a particle moving on the line $\mathbb{R}$. Assume that it is driven by a random velocity field $\varepsilon F(t)$ where ε is a small dimensionless parameter, F is stepwise constant

$$F(t) = \sum_{i=1}^{\infty} F_i \mathbf{1}_{|i-1,i|}(t),$$

and F_i are independent and identically distributed random variables that are bounded, $\mathbb{E}[F_i] = \bar{F}$ and $\mathbb{E}[(F_i - \bar{F})^2] = \sigma^2$. The position of the particle starting from 0 at time $t = 0$ is

$$X(t) = \varepsilon \int_0^t F(s)\,ds.$$

Clearly $X(t) \xrightarrow{\varepsilon \to 0} 0$. The problem consists in finding the adequate asymptotic, that is to say the scale of time which leads to a macroscopic motion of the particle.

Regime of the Law of Large Numbers. At the scale $t \to t/\varepsilon$, $X^\varepsilon(t) := X\left(\frac{t}{\varepsilon}\right)$ reads as:

$$
\begin{aligned}
X^\varepsilon(t) &= \varepsilon \int_0^{\frac{t}{\varepsilon}} F(s)\,ds \\
&= \varepsilon \left(\sum_{i=1}^{\left[\frac{t}{\varepsilon}\right]} F_i \right) + \varepsilon \int_{\left[\frac{t}{\varepsilon}\right]}^{\frac{t}{\varepsilon}} F(s)\,ds \\
&= \underset{\underset{t}{\downarrow}}{\varepsilon \left[\frac{t}{\varepsilon}\right]} \times \underset{\underset{\mathbb{E}[F] = \bar{F}}{\underset{\text{a.s.} \downarrow}{}}}{\frac{1}{\left[\frac{t}{\varepsilon}\right]} \left(\sum_{i=1}^{\left[\frac{t}{\varepsilon}\right]} F_i \right)} + \underset{\underset{0}{\underset{\text{a.s.} \downarrow}{}}}{\varepsilon \left(\frac{t}{\varepsilon} - \left[\frac{t}{\varepsilon}\right] \right) F_{\left[\frac{t}{\varepsilon}\right]}}.
\end{aligned}
$$

The convergence of $\frac{1}{\left[\frac{t}{\varepsilon}\right]}\left(\sum_{i=1}^{\left[\frac{t}{\varepsilon}\right]} F_i \right)$ is imposed by the Law of Large Numbers. Thus the motion of the particle is ballistic in the sense that it has constant effective velocity:

$$X^\varepsilon(t) \xrightarrow{\varepsilon \to 0} \bar{F}t.$$

However, in the case $\bar{F} = 0$, the random velocity field seems to have no effect, which means that we have to consider a different scaling.

Regime of the Central Limit Theorem $\bar{F} = 0$. At the scale $t \to t/\varepsilon^2$, $X^\varepsilon(t) = X\left(\frac{t}{\varepsilon^2}\right)$ reads as:

$$X^\varepsilon(t) = \varepsilon \int_0^{\frac{t}{\varepsilon^2}} F(s)\,ds$$

$$= \varepsilon \left(\sum_{i=1}^{\left[\frac{t}{\varepsilon^2}\right]} F_i \right) + \varepsilon \int_{\left[\frac{t}{\varepsilon^2}\right]}^{\frac{t}{\varepsilon^2}} F(s)\,ds$$

$$= \underbrace{\varepsilon \sqrt{\left[\frac{t}{\varepsilon^2}\right]}}_{\sqrt{t}} \times \underbrace{\frac{1}{\sqrt{\left[\frac{t}{\varepsilon^2}\right]}} \left(\sum_{i=1}^{\left[\frac{t}{\varepsilon^2}\right]} F_i \right)}_{\substack{\text{distribution} \downarrow \\ \mathcal{N}(0,\sigma^2)}} + \underbrace{\varepsilon \left(\frac{t}{\varepsilon^2} - \left[\frac{t}{\varepsilon^2}\right] \right) F_{\left[\frac{t}{\varepsilon^2}\right]}}_{\substack{\text{a.s.} \downarrow \\ 0}}.$$

The convergence of $\frac{1}{\sqrt{\left[\frac{t}{\varepsilon}\right]}} \left(\sum_{i=1}^{\left[\frac{t}{\varepsilon^2}\right]} F_i \right)$ is imposed by the Central Limit Theorem. $X^\varepsilon(t)$ converges in distribution as $\varepsilon \to 0$ to the Gaussian statistics $\mathcal{N}(0, \sigma^2 t)$. The motion of the particle in this regime is diffusive.

2.2 Stationary and ergodic processes

A stochastic process $(F(t))_{t\geq 0}$ is an application from some probability space to a functional space. This means that for any fixed time t the quantity $F(t)$ is a random variable with values in the state space E. E is a locally compact Polish space. We shall be often concerned with the case $E = \mathbb{R}^d$. Furthermore we shall only consider configurations where the functional space is either the set of the continuous functions $\mathcal{C}([0, \infty), E)$ equipped with the topology associated to the sup norm over the compact sets or the set of the càd-làg functions (right-continuous functions with left hand limits) equipped with the Skorohod topology. This means that the realizations of the random process are either continuous E-valued functions or càd-làg E-valued functions.

$(F(t))_{t\in\mathbb{R}^+}$ is a stationary stochastic process if the statistics of the process is invariant to a shift θ_t in the time origin:

$$\theta_t F \overset{\text{distribution}}{=} F$$

where the shift operator is

$$\theta_t F(\cdot) := F(t + \cdot).$$

It is a statistical steady state. Let us consider a stationary process such that $\mathbb{E}[|F(t)|] < \infty$. We set $\bar{F} = \mathbb{E}[F(t)]$. The ergodic theorem claims that the ensemble average can be substituted to the time average under the so-called ergodic hypothesis:

> If I is a subset of the phase space such that $\theta_t I = I$ a.s. for all t, then $\mathbb{P}(I) = 0$ or 1.

Theorem 2.1. *If F satisfies the ergodic hypothesis, then*

$$\frac{1}{T}\int_0^T F(t)\,dt \xrightarrow{T\to\infty} \bar{F} \quad \mathbb{P}\ a.s.$$

The ergodic hypothesis requires that the orbit $(F(t))_t$ visits all of phase space (difficult to state). This hypothesis can be made more transparent by looking at a non-ergodic process.

Example 2.2. Let F_1 and F_2 two ergodic processes, and denote $\bar{F}_j = \mathbb{E}[F_j(t)]$, $j = 1, 2$. Now flip a coin independently of F_1 and F_2, whose result is $\chi = 1$ with probability $1/2$ and 0 with probability $1/2$. Let $F(t) = \chi F_1(t) + (1 - \chi)F_2(t)$, which is a stationary process with mean $\bar{F} = \frac{1}{2}(\bar{F}_1 + \bar{F}_2)$. The time-averaged process satisfies

$$\frac{1}{T}\int_0^T F(t)\,dt = \chi\left(\frac{1}{T}\int_0^T F_1(t)\,dt\right) + (1 - \chi)\left(\frac{1}{T}\int_0^T F_2(t)\,dt\right)$$

$$\xrightarrow{T\to\infty} \chi\bar{F}_1 + (1 - \chi)\bar{F}_2$$

which is a random limit different from $\bar{F}$. This limit depends on χ because F has been trapped in a part of phase space.

2.3 Mean square theory

Let $(F(t))_{t\geq 0}$ be a stationary process, $\mathbb{E}[F^2] < \infty$. We introduce the autocorrelation function

$$R(\tau) = \mathbb{E}\left[(F(t) - \bar{F})(F(t + \tau) - \bar{F})\right].$$

By stationarity, R is an even function

$$R(-\tau) = \mathbb{E}\left[(F(t) - \bar{F})(F(t - \tau) - \bar{F})\right]$$
$$= \mathbb{E}\left[(F(t' + \tau) - \bar{F})(F(t') - \bar{F})\right] = R(\tau).$$

By Cauchy–Schwarz inequality, R reaches its maximum at 0:

$$R(\tau) \leq \mathbb{E}\left[(F(t) - \bar{F})^2\right]^{1/2}\mathbb{E}\left[(F(t + \tau) - \bar{F})^2\right]^{1/2} = R(0) = \mathrm{var}(F).$$

Proposition 2.3. *Assume that $\int_0^\infty \tau|R(\tau)|\,d\tau < \infty$. Let $S(T) = \frac{1}{T}\int_0^T F(t)\,dt$. Then*

$$\mathbb{E}\left[(S(T) - \bar{F})^2\right] \xrightarrow{T\to\infty} 0,$$

more exactly

$$T \, \mathbb{E}\big[(S(T) - \bar{F})^2\big] \xrightarrow{T \to \infty} 2 \int_0^\infty R(\tau) \, d\tau.$$

One should interpret the condition $\int_0^\infty \tau |R(\tau)| \, d\tau < \infty$ as "the autocorrelation function $R(\tau)$ decays to 0 sufficiently fast as $\tau \to \infty$." This hypothesis is a mean square version of mixing: $F(t)$ and $F(t + \tau)$ are approximatively independent for long time lags τ. Mixing substitutes for independence in the law of large numbers. In the case of piecewise constant processes:

$$F(s) = \sum_{k \in \mathbb{N}} f_k \mathbf{1}_{s \in [L_k, L_{k+1})}, \quad L_0 = 0, \quad L_k = \sum_{j=1}^{k} l_j$$

with independent and identically distributed random variables f_k, and independent exponential random variables l_k with mean 1 we have $R(\tau) = \mathrm{var}(f_1) \exp(-\tau)$.

Proof. The proof consists in a straightforward calculation.

$$\begin{aligned}
\mathbb{E}\big[(S(T) - \bar{F})^2\big] &= \mathbb{E}\left[\frac{1}{T^2} \int_0^T dt_1 \int_0^T dt_2 (F(t_1) - \bar{F})(F(t_2) - \bar{F})\right] \\
&= \frac{2}{T^2} \int_0^T dt_1 \int_0^{t_1} dt_2 \, R(t_1 - t_2) \quad \text{(by symmetry)} \\
&= \frac{2}{T^2} \int_0^T d\tau \int_0^{T-\tau} dh \, R(\tau) \quad \text{(with } \tau = t_1 - t_2, \; h = t_2) \\
&= \frac{2}{T^2} \int_0^T d\tau (T - \tau) R(\tau) \\
&= \frac{2}{T} \int_0^T R(\tau) \, d\tau - \frac{2}{T^2} \int_0^T \tau R(\tau) \, d\tau. \qquad \Box
\end{aligned}$$

Note that the $L^2(\mathbb{P})$ convergence implies convergence in probability as the limit is deterministic. Indeed, by Chebychev inequality, for any $\delta > 0$,

$$\mathbb{P}\left(|S(T) - \bar{F}| \geq \delta\right) \leq \frac{\mathbb{E}\big[(S(T) - \bar{F})^2\big]}{\delta^2} \xrightarrow{T \to \infty} 0.$$

2.4 The method of averaging

Let us revisit our toy model and consider a more general model for the velocity field. Let $0 < \varepsilon \ll 1$ be a small parameter and X^ε satisfies

$$\frac{dX^\varepsilon}{dt} = F\left(\frac{t}{\varepsilon}\right), \quad X^\varepsilon(0) = 0$$

where F is a stationary process with a decaying autocorrelation function such that $\int \tau |R(\tau)| d\tau < \infty$. $F(t)$ is a process on its own natural time scale. $F(t/\varepsilon)$ is the speeded-up process. The solution is $X^\varepsilon(t) = \int_0^t F(s/\varepsilon) ds = t\frac{1}{T}\int_0^T F(s) ds$ where $T = t/\varepsilon \to \infty$ as $\varepsilon \to 0$. So $X^\varepsilon(t) \to \bar{F}t$ as $\varepsilon \to 0$, or else $X^\varepsilon \to \bar{X}$ solution of:

$$\frac{d\bar{X}}{dt} = \bar{F}, \quad \bar{X}(t=0) = 0.$$

We can generalize this result to more general configurations.

Proposition 2.4 ([22, Khaminskii]). *Assume that, for each fixed value of $x \in \mathbb{R}^d$, $F(t, x)$ is a stochastic $\mathbb{R}^d$-valued process in t. Assume also that there exists a deterministic function $\bar{F}(x)$ such that*

$$\bar{F}(x) = \lim_{T \to \infty} \frac{1}{T} \int_{t_0}^{t_0+T} \mathbb{E}[F(t, x)] dt$$

with the limit independent of t_0. Let $\varepsilon > 0$ and X^ε be the solution of

$$\frac{dX^\varepsilon}{dt} = F\left(\frac{t}{\varepsilon}, X^\varepsilon\right), \quad X^\varepsilon(0) = 0.$$

Define $\bar{X}$ as the solution of

$$\frac{d\bar{X}}{dt} = \bar{F}(\bar{X}), \quad \bar{X}(0) = 0.$$

Then under wide hypotheses on F and $\bar{F}$, we have for any T:

$$\sup_{t \in [0,T]} \mathbb{E}\left[|X^\varepsilon(t) - \bar{X}(t)|\right] \xrightarrow{\varepsilon \to 0} 0.$$

Proof. The proof requires only elementary tools under the hypotheses:

1) F is stationary and $\mathbb{E}\left[\left|\frac{1}{T}\int_0^T F(t, x) dt - \bar{F}\right|\right] \xrightarrow{T \to \infty} 0$ (to check this, we can use the mean square theory since $\mathbb{E}[|Y|] \leq \sqrt{\mathbb{E}[Y^2]}$).

2) For any t $F(t, \cdot)$ and $\bar{F}(\cdot)$ are uniformly Lipschitz with a non-random Lipschitz constant c.

3) For any compact subset $K \subset \mathbb{R}^d$, $\sup_{t \in \mathbb{R}^+, x \in K} |F(t, x)| + |\bar{F}(x)| < \infty$.

We have

$$X^\varepsilon(t) = \int_0^t F\left(\frac{s}{\varepsilon}, X^\varepsilon(s)\right) ds, \quad \bar{X}(t) = \int_0^t \bar{F}(\bar{X}(s)) ds$$

so the difference reads

$$X^\varepsilon(t) - \bar{X}(t) = \int_0^t \left(F\left(\frac{s}{\varepsilon}, X^\varepsilon(s)\right) - F\left(\frac{s}{\varepsilon}, \bar{X}(s)\right)\right) ds + g^\varepsilon(t)$$

where $g^{\varepsilon}(t) := \int_0^t F\left(\frac{s}{\varepsilon}, \bar{X}(s)\right) - \bar{F}(\bar{X}(s))\, ds$. Taking the modulus,

$$|X^{\varepsilon}(t) - \bar{X}(t)| \leq \int_0^t \left| F\left(\frac{s}{\varepsilon}, X^{\varepsilon}(s)\right) - F\left(\frac{s}{\varepsilon}, \bar{X}(s)\right) \right| ds + |g^{\varepsilon}(t)|$$

$$\leq c \int_0^t |X^{\varepsilon}(s) - \bar{X}(s)|\, ds + |g^{\varepsilon}(t)|.$$

Taking the expectation and applying Gronwall's inequality,

$$\mathbb{E}\left[|X^{\varepsilon}(t) - \bar{X}(t)|\right] \leq e^{ct} \sup_{s \in [0,t]} \mathbb{E}\left[|g^{\varepsilon}(s)|\right].$$

It remains to show that the last term goes to 0 as $\varepsilon \to 0$. Let $\delta > 0$

$$g^{\varepsilon}(t) = \sum_{k=0}^{\lfloor t/\delta \rfloor - 1} \int_{k\delta}^{(k+1)\delta} \left(F\left(\frac{s}{\varepsilon}, \bar{X}(s)\right) - \bar{F}(\bar{X}(s)) \right) ds$$

$$+ \int_{\delta \lfloor t/\delta \rfloor}^{t} \left(F\left(\frac{s}{\varepsilon}, \bar{X}(s)\right) - \bar{F}(\bar{X}(s)) \right) ds.$$

Denote $M_T = \sup_{t \in [0,T]} |\bar{X}(t)|$. As $K_T = \sup_{x \in [-M_T, M_T], t \in \mathbb{R}^+} |F(t, x)| + |\bar{F}(x)|$ is finite, the last term of the right-hand side is bounded by $2 K_T \delta$. Furthermore F is Lipschitz, so that

$$\left| F\left(\frac{s}{\varepsilon}, \bar{X}(s)\right) - F\left(\frac{s}{\varepsilon}, \bar{X}(k\delta)\right) \right| \leq c \left| \bar{X}(s) - \bar{X}(k\delta) \right| \leq c K_T |s - k\delta|.$$

We have similarly:

$$\left| \bar{F}(\bar{X}(s)) - \bar{F}(\bar{X}(k\delta)) \right| \leq c K_T |s - k\delta|.$$

Thus

$$|g^{\varepsilon}(t)| \leq \left| \sum_{k=0}^{\lfloor t/\delta \rfloor - 1} \int_{k\delta}^{(k+1)\delta} \left(F\left(\frac{s}{\varepsilon}, \bar{X}(k\delta)\right) - \bar{F}(\bar{X}(k\delta)) \right) ds \right|$$

$$+ 2c K_T \sum_{k=0}^{\lfloor t/\delta \rfloor - 1} \int_{k\delta}^{(k+1)\delta} (s - k\delta)\, ds + 2 K_T \delta$$

$$\leq \varepsilon \sum_{k=0}^{\lfloor t/\delta \rfloor - 1} \left| \int_{k\delta/\varepsilon}^{(k+1)\delta/\varepsilon} \left(F(s, \bar{X}(k\delta)) - \bar{F}(\bar{X}(k\delta)) \right) ds \right| + 2 K_T (ct + 1)\delta.$$

Taking the expectation and the supremum:

$$\sup_{t \in [0,T]} \mathbb{E}[|g^{\varepsilon}(t)|] \leq \delta \sum_{k=0}^{\lfloor T/\delta \rfloor} \mathbb{E}\left[\left| \frac{\varepsilon}{\delta} \int_{k\delta/\varepsilon}^{(k+1)\delta/\varepsilon} \left(F(s, \bar{X}(k\delta)) - \bar{F}(\bar{X}(k\delta)) \right) ds \right| \right]$$

$$+ 2 K_T (cT + 1)\delta.$$

Taking the limit $\varepsilon \to 0$:

$$\limsup_{\varepsilon \to 0} \ \sup_{t \in [0,T]} \ \mathbb{E}[|g^{\varepsilon}(t)|] \leq 2K_T(cT + 1)\delta.$$

Letting $\delta \to 0$ completes the proof. ☐

3 Effective medium theory

This section is devoted to the computation of the effective velocity of an acoustic pulse traveling through a random medium. The acoustic pressure p and speed u satisfy the continuity and momentum equations

$$\rho \frac{\partial u}{\partial t} + \frac{\partial p}{\partial x} = 0 \tag{3.1}$$

$$\frac{\partial p}{\partial t} + \kappa \frac{\partial u}{\partial x} = 0 \tag{3.2}$$

where ρ is the material density and κ is the bulk modulus of the medium.

Assume that $\rho = \rho(x/\varepsilon)$ and $\kappa = \kappa(x/\varepsilon)$ are stationary random functions of position on spatial scale of order ε, $0 < \varepsilon \ll 1$. They are piecewise smooth and also uniformly bounded such that $\|\rho\|_{\infty} \leq C$, $\|\rho^{-1}\|_{\infty} \leq C$, $\|\kappa\|_{\infty} \leq C$, and $\|\kappa^{-1}\|_{\infty} \leq C$ a.s.

Assume conditions so that the system admits a solution, for instance a Dirichlet condition at $x = 0$ of the type $u(x = 0, t) = f(t)$ and $p(x = 0, t) = g(t)$ with $f, g \in L^2$. We also assume for simplicity that the Fourier transforms $\hat{f}$ and $\hat{g}$ decay faster than any exponential (say for instance that f and g are Gaussian pulses).

Note that the fluctuations in the sound speed are not assumed to be small. This corresponds to typical situations in acoustics and geophysics. The estimation of the vertical correlation length of the inhomogeneities in the lithosphere from well-log data is considered in [38]. They found that 2–3 m is a reasonable estimate of the correlation length of the fluctuations in sound speed. The typical pulse width is about 50 ms or, with a speed of 3 km/s, the typical wavelength is 150 m. So $\varepsilon = 10^{-2}$ is our framework.

We perform a Fourier analysis with respect to t. So we Fourier transform u and p

$$u(x, t) = \int \hat{u}(x, \omega)e^{-i\omega t} \, d\omega, \quad p(x, t) = \int \hat{p}(x, \omega)e^{-i\omega t} \, d\omega$$

so that we get a system of ordinary differential equations (ODE):

$$\frac{dX^{\varepsilon}}{dx} = F\left(\frac{x}{\varepsilon}, X^{\varepsilon}\right),$$

where

$$X^\varepsilon = \begin{pmatrix} \hat{p} \\ \hat{u} \end{pmatrix}, \quad F(x, X) = M(x)X, \quad M(x) = i\omega \begin{pmatrix} 0 & \rho(x) \\ \frac{1}{\kappa(x)} & 0 \end{pmatrix}.$$

Note first that a straightforward estimate shows that $|X^\varepsilon(\omega, z)| \le |X_0(\omega)| \exp(C\omega z)$ uniformly with respect to z. Apply the method of averaging. We get that $X^\varepsilon(x, \omega)$ converges in $L^1(\mathbb{P})$ to $\bar{X}(x, \omega)$ solution of

$$\frac{d\bar{X}}{dx} = \bar{M}\bar{X}, \quad \bar{M} = i\omega \begin{pmatrix} 0 & \bar{\rho} \\ \frac{1}{\bar{\kappa}} & 0 \end{pmatrix}, \quad \bar{\rho} = \mathbb{E}[\rho], \quad \bar{\kappa} = \left(\mathbb{E}[\kappa^{-1}]\right)^{-1}.$$

We now come back to the time domain. We introduce the deterministic "effective medium" with parameters $\bar{\rho}, \bar{\kappa}$ and the solution $(\bar{p}, \bar{u})$ of

$$\bar{\rho}\frac{\partial \bar{u}}{\partial t} + \frac{\partial \bar{p}}{\partial x} = 0$$

$$\frac{\partial \bar{p}}{\partial t} + \bar{\kappa}\frac{\partial \bar{u}}{\partial x} = 0.$$

The parameters are constant, so $\bar{u}$ satisfies the closed form equation $\frac{\partial^2 \bar{u}}{\partial t^2} - \bar{c}^2 \frac{\partial^2 \bar{u}}{\partial x^2} = 0$, which is the standard wave equation with the effective wave speed $\bar{c} = \sqrt{\bar{\kappa}/\bar{\rho}}$. We now compare $u^\varepsilon(x, t)$ with $\bar{u}(x, t)$:

$$\mathbb{E}\left[|u^\varepsilon(x, t) - \bar{u}(x, t)|\right] = \mathbb{E}\left[\left|\int e^{-i\omega t}(\hat{u}^\varepsilon(x, \omega) - \hat{\bar{u}}(x, \omega))\,d\omega\right|\right]$$

$$\le \int \mathbb{E}\left[|\hat{u}^\varepsilon(x, \omega) - \hat{\bar{u}}(x, \omega)|\right]d\omega.$$

The dominated convergence theorem then gives the convergence in $L^1(\mathbb{P})$ of u^ε to $\bar{u}$ in the time domain. Thus the effective speed of the acoustic wave $(p^\varepsilon, u^\varepsilon)$ as $\varepsilon \to 0$ is $\bar{c}$.

Example 3.1 (Bubbles in water). Air and water are characterized by the following parameters:

$\rho_a = 1.2\ 10^3$ g/m^3, $\kappa_a = 1.4\ 10^8$ g/s^2/m, $c_a = 340$ m/s;
$\rho_w = 1.0\ 10^6$ g/m^3, $\kappa_w = 2.0\ 10^{18}$ g/s^2/m, $c_w = 1425$ m/s.

If we consider a pulse whose frequency content is in the range 10 Hz–30 kHz, then the wavelengths lie in the range 1 cm–100 m. The bubble sizes are much smaller, so the effective medium theory can be applied. Let us denote by ϕ the volume fraction of air. The averaged density and bulk modulus are

$$\bar{\rho} = \mathbb{E}[\rho] = \phi\rho_a + (1 - \phi)\rho_w = \begin{cases} 9.9\ 10^5\ \text{g/m}^3 & \text{if } \phi = 1\% \\ 9\ 10^5\ \text{g/m}^3 & \text{if } \phi = 10\%, \end{cases}$$

$$\bar{\kappa} = \left(\mathbb{E}[\kappa^{-1}]\right)^{-1} = \left(\frac{\phi}{\kappa_a} + \frac{1 - \phi}{\kappa_w}\right)^{-1} = \begin{cases} 1.4\ 10^{10}\ \text{g/s}^2/\text{m} & \text{if } \phi = 1\% \\ 1.4\ 10^9\ \text{g/s}^2/\text{m} & \text{if } \phi = 10\%. \end{cases}$$

Accordingly $\bar{c} = 120$ m/s if $\phi = 1\%$ and $\bar{c} = 37$ m/s if $\phi = 10\,\%$.

The above example demonstrates that the average velocity may be much smaller than the minimum of the velocities of the medium components. However it cannot happen in such a configuration that the velocity be larger than the maximum (or the essential supremum) of the velocities of the medium components. Indeed,

$$\mathbb{E}\big[c^{-1}\big] = \mathbb{E}\big[\kappa^{-1/2}\rho^{1/2}\big] \le \mathbb{E}\big[\kappa^{-1}\big]^{1/2}\mathbb{E}[\rho]^{1/2} = \bar{c}^{-1}.$$

Thus $\bar{c} \le \mathbb{E}\big[c^{-1}\big]^{-1} \le \mathrm{ess\ sup}(c)$.

4 Markov processes

4.1 Definition and main properties

A stochastic process $(X_t)_{t\ge 0}$ with state space E is a Markov process if $\forall 0 \le s < t$ and $B \in \mathcal{B}(E)$ (the σ-algebra of Borel sets of E)

$$\mathbb{P}(X_t \in B | X_u, u \le s) = \mathbb{P}(X_t \in B | X_s)$$

"the state X_s at time s contains all relevant information for calculating probabilities of future events". The rigorous definition needs σ-algebras of measurable sets. This is a generalization to the stochastic case of the dynamical deterministic systems without memory of the type $\frac{dx}{dt} = f(t, x(t))$.

The distribution of X_t starting from x at time s is the transition probability

$$\mathbb{P}(X_t \in B | X_s = x) = P(s, x; t, B) = \int_{y \in B} P(s, x; t, dy).$$

Definition 4.1. A transition probability P is a function from $\mathbb{R}^+ \times E \times \mathbb{R}^+ \times \mathcal{B}(E)$ such that

1) $P(s, x; t, A)$ is measurable in x for fixed $s \in \mathbb{R}^+, t \in \mathbb{R}^+, A \in \mathcal{B}(E)$,

2) $P(s, x; t, A)$ is a probability measure in A for fixed $s \in \mathbb{R}^+, t \in \mathbb{R}^+, x \in E$,

3) P satisfies the Chapman–Kolmogorov equation:

$$P(s, x; t, A) = \int_E P(s, x; \tau, dz)P(\tau, z; t, A) \quad \forall 0 \le s < \tau < t.$$

Note that the Chapman–Kolmogorov equation can be deduced heuristically from a disintegration formula:

$$\mathbb{P}(X_t \in A | X_s = x) = \int_E \mathbb{P}(X_t \in A | X_s = x, X_\tau = z)\mathbb{P}(X_\tau \in dz | X_s = x)$$

and the application of the Markov property.

Note also that a Markov process is temporally homogeneous if the transition probability depends only on $t - s$: $P(s, x; t, A) = P(0, x; t - s, A)$.

We can now define the 2-parameter family of operators defined on the space of measurable bounded functions $L^\infty(E)$:

$$T_{s,t} f(x) = \mathbb{E}[f(X_t)|X_s = x] = \int_E P(s, x; t, dy) f(y).$$

Proposition 4.2. 1) $T_{t,t} = I_d$.

2) $\forall s \leq \tau \leq t \ T_{s,t} = T_{s,\tau} T_{\tau,t}$.

3) $T_{s,t}$ is a contraction $\|T_{s,t} f\|_\infty \leq \|f\|_\infty$.

Proof. The second point follows from the Chapman–Kolmogorov relations, and the third one:

$$|T_{s,t} f(x)| \leq \int_E p(s, x; t, dy)\|f\|_\infty = \|f\|_\infty. \qquad \square$$

We shall consider Feller processes: for any continuous bounded function f such that $\lim_{|x|\to\infty} f(x)$ exists and equals 0 and for any $t > 0$ the function $(h, x) \mapsto T_t^{t+h} f(x)$ is continuous. This continuity property goes from the family operators $(T_{s,t})$ to the process itself. The exact assertion claims that this is true for a modification of the process. A stochastic process $\tilde{X}$ is called a modification of a process X if $\mathbb{P}(X_t = \tilde{X}_t) = 1$ for all t.

Proposition 4.3. *Let (X_t) be a Feller Markov process. Then (X_t) has a modification whose realizations are càd-làg functions.*

From now on we shall always consider such a modification of the Markov process. The generator of a Markov process is the operator

$$Q_t := \lim_{s \nearrow t} \frac{T_s^t - I_d}{t - s}.$$

It is defined on a subset of $L^\infty(E)$.

Proposition 4.4. *If $u(s, \cdot, \cdot) \in \mathrm{Dom}(Q_s)$, then the function $u(s, t, x) := T_{s,t} f(x)$ satisfies the Kolmogorov backward equation (KBE)*

$$\frac{\partial u}{\partial s} + Q_s u = 0, \quad s < t, \ u(s = t, t, x) = f(x). \tag{4.1}$$

Proof.

$$Q_s u = \lim_{\delta \to 0} \frac{T_{s-\delta}^s - I_d}{\delta} u(s, t, x) = -\lim_{\delta \to 0} \frac{I_d - T_{s-\delta}^s}{\delta} T_s^t f(x)$$

$$= -\lim_{\delta \to 0} \frac{T_s^t f - T_{s-\delta}^t f}{\delta} = -\lim_{\delta \to 0} \frac{u(t, s, x) - u(t, s - \delta, x)}{\delta} = -\frac{\partial u}{\partial s}. \qquad \square$$

Application: ODE driven by a Feller Markov process. Let q be a Feller process with generator Q_t and X be the solution of:

$$\frac{dX}{dt} = F(q(t), X(t))$$

where F is Lipschitz continuous and bounded. Then $Y = (q, X)$ is a Markov process with generator:

$$\mathcal{L} = Q_t + \sum_{j=1}^{d} F_j(q, X) \frac{\partial}{\partial X_j}.$$

Proof. The fact that Y is Markov is straightforward. Let f be a smooth bounded test function.

$$\mathbb{E}\left[f(Y(t+h)) \mid Y(t) = (X, q)\right] - f(y) = A_h + B_h \qquad (4.2)$$

where

$$A_h = \mathbb{E}\left[f(X(t+h), q(t+h)) - f(X(t), q(t+h)) \mid Y(t) = (X, q)\right]$$
$$B_h = \mathbb{E}\left[f(X(t), q(t+h)) - f(X(t), q(t)) \mid Y(t) = (X, q)\right].$$

On the one hand

$$A_h = \mathbb{E}\left[\nabla f(X(t), q(t+h)).(X(t+h) - X(t))\right.$$
$$\left. + O(|X(t+h) - X(t)|^2) \mid Y(t) = (X, q)\right]$$

$$= \mathbb{E}\left[\nabla f(X(t), q(t+h)).F(X(t)) \mid Y(t) = (X, q)\right] h + O(h^2)$$

which implies (q is assumed to be Feller):

$$\frac{A_h}{h} \xrightarrow{h \to 0} \mathbb{E}\left[\nabla f(X(t), q(t)).F(X(t)) \mid Y(t) = (X, q)\right] = \nabla f(X, q).F(X).$$
$$(4.3)$$

On the other hand

$$\frac{B_h}{h} = \frac{1}{h}\mathbb{E}\left[f(X, q(t+h)) - f(X, q(t)) \mid q(t) = q\right] \xrightarrow{h \to 0} Q_t f(X, q). \quad (4.4)$$

Substituting (4.3) and (4.4) into (4.2) yields the result. $\square$

4.2 Homogeneous Feller processes

In this section we consider a temporally homogeneous Markov process. Its transition probability $P(s, x; t, A)$ is then denoted by $P_{t-s}(x, A)$, and the family of operators $T_{s,t}$ is then of the form $T_{s,t} = T_{t-s}$:

$$T_t f(x) = \int_E P_t(x, dy) f(y).$$

It is a semi-group as it satisfies $T_{t+s} = T_t T_s$. It is not a group as T_t does not possess an inverse. The Feller property then claims that T_t is strongly continuous.

The distribution $\boldsymbol{P}_x$ of the Markov process starting from $x \in E$ is described by the probability transition. Indeed, by the Chapman–Kolmogorov equation, for any $n \in \mathbb{N}^*$, for any $0 < t_1 < t_2 < \cdots < t_n < \infty$ and for any $A_1, \ldots, A_n \in \mathcal{B}(E)$, we have:

$$\boldsymbol{P}_x(X_{t_1} \in A_1, \ldots, X_{t_n} \in A_n)$$
$$= \int_{A_1} \cdots \int_{A_n} P_{t_1}(x, dx_1) P_{t_2-t_1}(x_1, dx_2) \ldots P_{t_n-t_{n-1}}(x_{n-1}, dx_n).$$

We shall denote by $\boldsymbol{E}_x$ the expectation with respect to $\boldsymbol{P}_x$. We have in particular $\boldsymbol{E}_x[f(X_t)] = T_t f(x)$.

The Markov processes have been extensively studied and classified. This classification is based upon the notions of recurrence and transience. From this classification simple conditions for the ergodicity of the process can be deduced.

The main hypothesis that will insure most of the forthcoming results is that the transition probability has a positive, continuous density function:

Hypothesis D. *There exists a Borel measure μ on E supported by E, and a strictly positive function $p_t(x, y)$ continuous in $(t, x, y) \in \mathbb{R}^{+*} \times E^2$ such that the transition probability $P_t(x, dy)$ equals $p_t(x, y)\mu(dy)$.*

Let $L^\infty(E)$ be the set of all bounded measurable functions on E and let $\mathcal{C}_b(E)$ be the set of all bounded continuous functions on E. For each $t > 0$, T_t maps $L^\infty(E)$ into $\mathcal{C}_b(E)$. Indeed if $f \in L^\infty(E)$ satisfies $0 \leq f \leq 1$, both $T_t f$ and $T_t(1 - f)$ are lower semicontinuous and their sum is constant 1. Then $T_t f$ has to be continuous. This can be generalized to any $f \in L^\infty(E)$. The semigroup with this property is called a strong Feller semigroup.

4.2.1 Recurrent and transient properties. The Feller process is called recurrent (in the sense of Harris) if

$$\boldsymbol{P}_x \left(\int_0^\infty \mathbf{1}_A(X_t) \, dt = \infty \right) = 1$$

for every $x \in E$ and $A \in \mathcal{B}(E)$ such that $\mu(A) > 0$. This means that the time spent by the process in any subset is infinite, or else that it comes back an infinite number of times in any subset.

The Feller process is called transient if

$$\sup_{x \in E} \boldsymbol{E}_x \left[\int_0^\infty \mathbf{1}_A(X_t) \, dt \right] < \infty$$

for every compact subset of E. This means that the process spends only a finite time in any compact subset, so that it goes to infinity. The classification result holds as follows.

Proposition 4.5. *A Feller process that satisfies Hypothesis D is either recurrent or transient.*

The proof can be found in |33|. It consists in 1) introducing a suitable Markov process with discrete time parameters, 2) showing a similar transient-recurrent dichotomy for this process, and 3) applying this result to the Feller process.

4.2.2 Invariant measures. Let π be a Borel measure on E. It is called an invariant measure of the Feller semigroup (T_t) if

$$\int_E T_t f(x)\pi(dx) = \int_E f(x)\pi(dx)$$

for every $t \geq 0$ and for any nonnegative functions with compact support. The following proposition (whose proof can be found for instance in |26|) shows that a recurrent process possesses a unique invariant measure.

Proposition 4.6. *Let (X_t) be a Feller process satisfying Hypothesis D. If it is recurrent, then there exists an invariant measure π which is unique up to a multiplicative constant and mutually absolutely continuous with respect to the reference Borel measure μ.*

In the case where the state space is compact, the process cannot be transient, hence it is recurrent and possesses an invariant measure, which has to be a bounded Borel measure.

Corollary 4.7. *Let (X_t) be a Feller process with compact state space satisfying Hypothesis D. Then it is recurrent and has a unique invariant probability measure.*

4.2.3 Ergodic properties. Let $f \in L^\infty(E)$. It is called T_t-invariant if $T_t f(x) = f(x)$ for any t and x. We can characterize the ergodic hypothesis in terms of T_t-invariant functions.

Proposition 4.8. *A Feller process is ergodic if every bounded measurable T_t-invariant function is a constant function. Conversely if the Feller process is ergodic, then every bounded continuous T_t-invariant function is a constant function.*

The ergodic theorems that we are going to state are based on the following proposition.

Proposition 4.9. *Let (X_t) be a recurrent Feller process satisfying Hypothesis D. For every pair of probabilities π_1 and π_2, we have*

$$\lim_{t \to \infty} \|(\pi_1 - \pi_2)P_t\| = 0$$

where $\pi_1 P_t(\cdot) = \int_E \pi_1(x)P_t(x, \cdot)$ and $\| \cdot \|$ is here the norm of the total variations.

The proposition means that the process forgets its initial distribution as $t \to \infty$. Applying the proposition with $\pi_1 = \delta_x$ and $\pi_2 =$ the invariant measure of the process is the key to the proof of the ergodic theorem for Markov processes:

Proposition 4.10. *Let* (X_t) *be a Feller process satisfying Hypothesis D.*

1) *If it is recurrent and has an invariant probability measure* π, *then for any* $f \in L^\infty(E)$ *and for any* $x \in E$,

$$T_t f(x) \xrightarrow{\ t \to \infty\ } \int_E f(y) \pi(dy).$$

2) *If it is recurrent and has an infinite invariant measure, then for any* $f \in L^\infty(E)$ *such that* $\int_E |f(y)| \pi(dy) < \infty$ *and for any* $x \in E$:

$$T_t f(x) \xrightarrow{\ t \to \infty\ } 0.$$

3) *If it is transient, then for any* $f \in L^\infty(E)$ *with compact support and for any* x:

$$T_t f(x) \xrightarrow{\ t \to \infty\ } 0.$$

The ergodic theorem has also a pathwise version:

Proposition 4.11. *Let* (X_t) *be a Feller process satisfying Hypothesis D.*

1) *If it is recurrent and has an invariant probability measure* π, *then for any* $f \in L^\infty(E)$ *and for any* $x \in E$,

$$\frac{1}{t} \int_0^t f(X_t) \xrightarrow{\ t \to \infty\ } \int_E f(y) \pi(dy) \quad \boldsymbol{P}_x \text{ almost surely.}$$

2) *If it is recurrent and has an infinite invariant measure, then for any* $f \in L^\infty(E)$ *such that* $\int_E |f(y)| \pi(dy) < \infty$ *and for any* $x \in E$:

$$f(X_t) \xrightarrow{\ t \to \infty\ } 0 \quad \text{in } L^1(\boldsymbol{P}_x).$$

3) *If it is transient, then for any* $f \in L^\infty(E)$ *with compact support and for any* x:

$$f(X_t) \xrightarrow{\ t \to \infty\ } 0 \quad \boldsymbol{P}_x \text{ almost surely.}$$

4.2.4 Resolvent equations and potential kernels. We consider a Feller process satisfying Hypothesis D. We set, for $\alpha > 0$,

$$u_\alpha(x, y) = \int_0^\infty e^{-\alpha t} p_t(x, y) \, dt$$

which is a strictly lower semicontinuous function. The family of operators $(U_\alpha)_{\alpha > 0}$

$$U_\alpha(x) = \int_E u_\alpha(x, y) f(y) \mu(dy) = \int_0^\infty e^{-\alpha t} T_t f(x) \, dt$$

is called the resolvent of the semigroup (T_t). It satisfies the resolvent equation:

$$(\alpha - \beta)U_\alpha U_\beta f + U_\alpha f - U_\beta f = 0 \qquad (4.5)$$

for any $\alpha, \beta > 0$.

A kernel U (i.e. a family of Borel measures $\{U(x, \cdot), x \in E\}$ such that $U(x, A)$ is measurable with respect to x for all $A \in \mathcal{B}(E)$) is called a potential kernel if it satisfies:

$$(I - \alpha U_\alpha)Uf = U_\alpha f \qquad (4.6)$$

for every $\alpha > 0$ and f such that $Uf \in \mathcal{C}_b(E)$. The function Uf is called a potential of f. Note that Eq. (4.6) is the limit form of Eq. (4.5) as $\beta \to 0$.

If the process is transient, then

$$U(x, A) := \int_0^\infty P_t(x, A) \, dt \qquad (4.7)$$

is a potential kernel. Indeed if $f \in L^\infty(E)$ with compact support, $Uf \in \mathcal{C}_b(E)$ and satisfies $Uf = \lim_{\beta \to 0} U_\beta f$. Therefore it satisfies (4.6) since (U_α) satisfies the resolvent equation (4.5).

If the process is recurrent and has invariant measure π, then (4.7) diverges as soon as $\pi(A) > 0$. However it is possible to construct a kernel W such that $Wf \in \mathcal{C}_b(E)$ and satisfies (4.6) if $f \in L^\infty(E)$ with compact support satisfying $\int_E f(y)\pi(dy) = 0$. Such a kernel is called a recurrent potential kernel.

Proposition 4.12. *Let (X_t) be a recurrent Feller process satisfying Hypothesis D. There exists a recurrent potential W. Assume further that the process has an invariant probability measure π. If $f \in L^\infty(E)$ with compact support satisfies $\int_E f(y)\pi(dy) = 0$, then $\int_0^t T_s f(x)ds$ is bounded in (t, x) and*

$$\int_0^t T_s f(x) \, ds \xrightarrow{t \to \infty} Wf(x) - \int_E Wf(y)\pi(dy).$$

Note that in the recurrent case with invariant probability measure, the recurrent potential kernel exists and is unique if one adds the condition that $\int_E Wf(y)d\pi(dy) = 0$ for all $f \in L^\infty(E)$ with compact support satisfying $\int_E f(y)\pi(dy) = 0$. Furthermore, if Wf belongs to the domain of the infinitesimal generator of the semigroup (T_t), then the potential kernel satisfies $QWf = (\alpha - U_\alpha^{-1})Wf = -f$. We can then state the important following corollary.

Corollary 4.13. *Let (X_t) be a recurrent Feller process with an invariant probability measure π satisfying Hypothesis D. If $f \in L^\infty(E)$ with compact support satisfies $\int_E f(y)\pi(dy) = 0$, and if Wf belongs to the domain of the infinitesimal generator of the semigroup (T_t), then Wf is a solution of the Poisson equation $Qu = -f$.*

4.3 Diffusion Markov processes

Definition 4.14. Let P be a transition probability. It is associated to a diffusion process if:

1) $\forall x \in \mathbb{R}^d, \forall \varepsilon > 0, \int_{|y-x|>\varepsilon} P(s, x; t, dy) = o(t - s)$ uniformly over $s < t$,

2) $\forall x \in \mathbb{R}^d, \forall \varepsilon > 0, \int_{|y-x|\leq\varepsilon}(y_i - x_i)P(s, x; t, dy) = b_i(s, x)(t - s) + o(t - s)$ for $i = 1, \ldots, d$ uniformly over $s < t$,

3) $\forall x \in \mathbb{R}^d, \forall \varepsilon > 0, \int_{|y-x|\leq\varepsilon}(y_i - x_i)(y_j - x_j)P(s, x; t, dy) = a_{ij}(s, x)(t - s) + o(t - s)$ for $i, j = 1, \ldots, d$, uniformly over $s < t$.

The functions b_i characterize the drift of the process, while a_{ij} describe the diffusive properties of the diffusion process. In the following we shall restrict ourselves to homogeneous diffusion processes, whose diffusive matrix a and drift b do not depend on time. We are going to see that these functions completely characterize a diffusion Markov process with some more technical hypotheses.

We shall assume the following hypotheses:

(H) $\begin{cases} a_{ij} \text{ are of class } \mathcal{C}^2 \text{ with bounded derivatives,} \\[6pt] b_i \text{ are of class } \mathcal{C}^1 \text{ with bounded derivatives,} \\[6pt] a \text{ satisfies the strong ellipticity condition:} \\ \text{There exists some } \gamma > 0 \text{ such that, for any } (t, x), \\ \sum_{ij} a_{ij}(x)\xi_i\xi_j \geq \gamma \sum_i \xi_i^2. \end{cases}$

We introduce the second-order differential operator L:

$$Lf(x) = \sum_{i,j=1}^d a_{ij}(x)\frac{\partial^2 f(x)}{\partial x_i \partial x_j} + \sum_{i=1}^d b_i(x)\frac{\partial f(x)}{\partial x_i}.$$

Proposition 4.15. *Under Hypotheses* (H):

1) *There exists a unique Green's function* $p_t(x, y)$ *from* $\mathbb{R}^d \times \mathbb{R}^+ \times \mathbb{R}^d$ *to* $\mathbb{R}$ *such that*

$p_t(x, y) > 0 \ \forall t > 0, \ x, y \in \mathbb{R}^d,$

p is continuous on $\mathbb{R}^d \times \mathbb{R}^{+*} \times \mathbb{R}^d,$

p is $\mathcal{C}^2$ *in x and* $\mathcal{C}^1$ *in t,*

as a function of t and x, p satisfies $\frac{\partial p}{\partial t} = Lp,$

$\forall x, \int_{\mathbb{R}^d} p_t(x, y)f(y)dy \rightarrow f(x)$ *as* $t \rightarrow 0^+$ *for any continuous and bounded function* f.

2) *There exists a unique positive function $\bar{p} \in C^1(\mathbb{R}^d, \mathbb{R})$ such that $L^*\bar{p} \equiv 0$ up to a multiplicative constant. If $\bar{p}$ has finite mass, then we choose the normalization $\int_{\mathbb{R}^d} \bar{p}(y)dy = 1$.*

We refer to [16] for the proof. This proposition is obtained by means of PDE tools. The first point is the key for the proof of the following proposition that describes a class of Markov processes satisfying Hypothesis D.

Proposition 4.16. *Under Hypotheses (H), there exists a unique diffusion Markov process with drift b and diffusive matrix a. It is Feller and it satisfies hypothesis D. It has continuous sample paths. L is its infinitesimal generator. The Green's function p is the transition probability density*

$$P_t(x, A) = \int_A p_t(x, y)\, dy$$

which satisfies the KBE as a function of t and x:

$$\frac{\partial p}{\partial t} = Lp, \quad p_{t=0}(x, y) = \delta(x - y),$$

and the Kolmogorov forward equation (KFE) as a function of t and y:

$$\frac{\partial p}{\partial t} = L^*p, \quad p_{t=0}(x, y) = \delta(x - y),$$

where L^* is the adjoint operator of L

$$L^* f(x) = \sum_{i,j=1}^{d} \frac{\partial^2}{\partial x_i \partial x_j} \left(a_{ij}(x) f(x) \right) - \sum_{i=1}^{d} \frac{\partial}{\partial x_i} \left(b_i(x) f(x) \right).$$

Note that the KBE is the same as Eq. (4.1), as the probability transition reads $P(s, x; t, A) = P(0, x; t - s, A)$ so that $\frac{\partial p}{\partial s} = -\frac{\partial p}{\partial t}$. The KFE is also known as the Fokker–Planck equation.

Furthermore the KBE and KFE are not restricted to the diffusion processes. We can actually state in great generality for a Markov process (X_t) with infinitesimal generator Q that, if dom(Q) is a subset of L^∞ sufficiently large and if the probability density function p exists and is sufficiently smooth, then it satisfies both the KBE and the KFE. Indeed, from Proposition 4.4, if $u \in$ dom(Q) and p is smooth enough, we have

$$\int \frac{\partial p(s, x; t, y)}{\partial s} f(y)\, dy = \frac{\partial u}{\partial s} = -Qu = -\int Qp(s, x; t, y)f(y)\, dy$$

which proves the KBE for p smooth enough. Second, by the Chapman–Kolmogorov equation,

$$0 = \frac{\partial}{\partial \tau} p(s, x; t, y)$$

$$= \int \frac{\partial p(s, x; \tau, z)}{\partial \tau} p(\tau, z; t, y) \, dz + \int p(s, x; \tau, z) \frac{\partial p(\tau, z; t, y)}{\partial \tau} \, dz$$

$$= \int \frac{\partial p(s, x; \tau, z)}{\partial \tau} p(\tau, z; t, y) \, dz - \int p(s, x; \tau, z) Q_\tau p(\tau, z; t, y) \, dz$$

$$= \int \frac{\partial p(s, x; \tau, z)}{\partial \tau} p(\tau, z; t, y) \, dz - \int Q_\tau^* p(s, x; \tau, z) p(\tau, z; t, y) \, dz$$

which proves the KFE if we let $\tau \nearrow t$ and use $p(t, y; \tau, z) \to \delta(y - z)$.

A diffusion process can be either transient or recurrent. A criterion that insures recurrence is that there exists $K > 0, \alpha \geq -1, r > 0$ such that

$$\sum_{j=1}^{d} b_j(x) \frac{x_j}{|x|} \leq -r|x|^\alpha \quad \forall \, |x| \geq K$$

(r should be large enough in the case $\alpha = -1$). In such a case the drift b plays the role of a trapping force that pushes the process to the origin [37].

The following proposition expresses the Fredholm alternative for the operator L in the ergodic case. It can be seen as a consequence of Proposition 4.12 and Corollary 4.13 in the framework of diffusion processes.

Proposition 4.17. *Under Hypotheses* (H), *if the invariant measure has finite mass, then the diffusion process is ergodic. If moreover $f \in C^2$ with compact support satisfies $\int_{\mathbb{R}^d} f(y)\bar{p}(y)dy = 0$, then there exists a unique function χ which satisfies $L\chi = f$ and the centering condition $\int_{\mathbb{R}^d} \chi(y)\bar{p}(y)dy \equiv 0$. It is given by $\int_0^\infty T_s f(x)ds = \int_0^\infty \int_{\mathbb{R}^d} p_s(x, y)f(y)dyds$.*

Example 4.18 (The Brownian motion). The d-dimensional Brownian motion is the $\mathbb{R}^d$-valued homogeneous Markov process with infinitesimal generator:

$$Q = \frac{1}{2}\Delta$$

where Δ is the standard Laplacian operator. The probability transition density

$$p(0, x; t, y)$$

has the Gaussian density with mean x and covariance matrix $t I_d$:

$$p(0, x; t, y) = \frac{1}{(2\pi t)^{d/2}} \exp\left(-\frac{|y - x|^2}{2t}\right).$$

An explicit calculation demonstrates that the Brownian motion is recurrent in dimensions 1 and 2 and transient in higher dimensions. The Brownian motion possesses an invariant measure which is simply the Lebesgue measure over $\mathbb{R}^d$ which has infinite mass. Whatever the dimension is, the Brownian motion is not ergodic, it escapes to infinity. It will not be a suitable process for describing a stationary ergodic medium. However if we add a trapping potential the Brownian motion becomes ergodic, as shown in the next example.

Example 4.19 (The Ornstein–Uhlenbeck process). It is an $\mathbb{R}$-valued homogeneous Markov process defined as the solution of the stochastic differential equation

$$dX_t = -\lambda X_t + dW_t$$

that admits the closed form expression

$$X_t = X_0 e^{-\lambda t} + \int_0^t e^{-\lambda(t-s)} dW_s,$$

where W is a standard one-dimensional Brownian motion. It describes the evolution of the position of a diffusive particle trapped in a quadratic potential. It is the homogeneous Markov process with infinitesimal generator:

$$Q = \frac{1}{2} \frac{\partial^2}{\partial x^2} - \lambda x \frac{\partial}{\partial x}.$$

The probability transition $p(0, x; t, \cdot)$ has a Gaussian density with mean $x e^{-\lambda t}$ and variance $\sigma^2(t)$:

$$p(0, x; t, y) = \frac{1}{\sqrt{2\pi \sigma(t)^2}} \exp\left(-\frac{(y - x e^{-\lambda t})^2}{2\sigma^2(t)}\right), \quad \sigma^2(t) = \frac{1 - e^{-2\lambda t}}{2\lambda}.$$

It is a recurrent ergodic process whose invariant probability measure has a density with respect to the Lebesgue measure:

$$\bar{p}(y) = \sqrt{\frac{\lambda}{\pi}} \exp\left(-\lambda y^2\right). \tag{4.8}$$

Note that it may be more comfortable in some circumstances to deal with a process with compact state space. For instance the process $Y_t = \arctan(X_t)$ with (X_t) an Ornstein–Uhlenbeck process is one of the models that can be used to describe the fluctuations of the parameters of a random medium.

4.4 The Poisson equation and the Fredholm alternative

We consider in this section an ergodic Feller Markov process with infinitesimal generator Q. The probability transitions converge to the invariant probability measure by the ergodic theorem. The resolution of the Poisson equation $Qu = f$ requires fast enough mixing. A set of hypotheses for rapid convergence is stated in the following proposition due to Doeblin:

Proposition 4.20. *Assume that the Markov process has a compact state space E, it is Feller, and there exists $t_0 > 0$, $c > 0$ and a probability μ over E such that $P_t(x, A) \geq c\mu(A)$ for all $t \geq t_0$, $x \in E$, $A \in \mathcal{B}(E)$. Then there exists a unique invariant probability $\bar{p}$ and two positive numbers $c_1 > 0$ and $\delta > 0$ such that*

$$\sup_{x \in E} \sup_{A \in \mathcal{B}(E)} |P_t(x, A) - \bar{p}(A)| \leq c_1 e^{-\delta t}.$$

A Feller process with compact state space E satisfying Hypothesis D possesses nice mixing properties. Indeed, if $t_0 > 0$, then $\delta := \inf_{x,y \in E} p(0, x; t_0, y)$ is positive as E is compact and p is continuous. Denoting by μ the uniform distribution over E we have $P_{t_0}(x, A) \geq \delta\mu(A)$ for all $A \in \mathcal{B}(E)$. This property also holds true for any time $t_0 + t$, $t \geq 0$, as the Chapman–Kolmogorov relation implies $P_{t+t_0}(x, A) = \int_E P_t(x, dz) P_{t_0}(z, A) \geq \delta \int_E \mu(A) P_t(x, dz) = \delta\mu(A)$. This proves the following corollary:

Corollary 4.21. *If a Feller process with compact state space satisfies Hypothesis D, then it fulfills the hypotheses and conclusion of Proposition* 4.20.

We end up the section by revisiting the problem of the Poisson equation in terms of the Fredholm alternative. We consider an ergodic Feller Markov process satisfying the Fredholm alternative. We first investigate the null spaces of the generator Q. Considering $T_t 1 = 1$, we have $Q1 = 0$, so that $1 \in \text{Null}(Q)$. As a consequence $\text{Null}(Q^*)$ is at least one-dimensional. As the process is ergodic $\text{Null}(Q^*)$ is exactly one-dimensional. In other words there exists a unique invariant probability measure which satisfies $Q^*\bar{p} = 0$. In such conditions the probability transition converges to $\bar{p}$ as $t \to \infty$. The spectrum of Q^* gives the rate of forgetfulness, i.e. the mixing rate. For instance the existence of a spectral gap

$$\inf_{f.\int_E f d\bar{p}=0} \frac{-\int_E f Q f d\bar{p}}{\int_E f^2 d\bar{p}} > 0$$

insures an exponential convergence of $P_t(x, \cdot)$ to $\bar{p}$. We now investigate the solutions of the Poisson equation $Qu = f$. Of course Q is not invertible since it has a nontrivial null space $\{1\}$. $\text{Null}(Q^*)$ has dimension 1 and is generated by the invariant probability measure $\bar{p}$. By the Fredholm alternative, the Poisson equation admits a solution if f satisfies the orthogonality condition $f \perp \text{Null}(Q^*)$ which means that f has zero mean $\mathbb{E}[f(X_0)] = 0$ where $\mathbb{E}$ is the expectation with respect to the distribution of the Markov process starting with the invariant measure $\bar{p}$:

$$\mathbb{E}[f(X_t)] = \int_E \bar{p}(dx)\mathbb{E}_x[f(X_t)].$$

In such a case, a particular solution of the Poisson equation $Qu = f$ is

$$u_0(x) = -\int_0^\infty ds\, T_s f(x).$$

Remember that the following expressions are equivalent:

$$T_s f(x) = \int_E P_s(x, dy) f(y) = \mathbb{E}_x[f(X_s)] = \mathbb{E}[f(X_s)|X_0 = x].$$

The convergence of the integral needs fast enough mixing. Also we have formally

$$Q u_0 = -\int_0^\infty ds\, Q e^{sQ} f = -\int_0^\infty ds\, \frac{de^{sQ}}{ds} f = -\left[e^{sQ} f\right]_0^\infty = f - \mathbb{E}[f(X_0)] = f.$$

Note that another solution of the Poisson equation belongs to Null(Q) and is therefore a constant. Here we have $\mathbb{E}[u_0(X_0)] = 0$ as $\mathbb{E}[f(X_s)] = \mathbb{E}[f(X_0)] = 0$, so that we can claim that $-\int_0^\infty ds\, T_s : \mathcal{D} \to \mathcal{D}$ is the inverse of Q restricted to $\mathcal{D} = (\text{Null}(Q^*))^\perp$.

4.5 Diffusion-approximation for Markov processes

Proposition 4.22. *Let us consider the system*

$$\frac{dX^\varepsilon}{dt}(t) = \frac{1}{\varepsilon} F\left(X^\varepsilon(t), q\left(\frac{t}{\varepsilon^2}\right), \frac{t}{\varepsilon}\right), \quad X^\varepsilon(0) = x_0 \in \mathbb{R}^d. \qquad (4.9)$$

Assume that

1) q is a Markov, stationary, ergodic process on a compact space with generator Q, satisfying the Fredholm alternative,

2) F satisfies the centering condition: $\mathbb{E}[F(x, q(0), \tau)] = 0$ for all x and τ where $\mathbb{E}[\cdot]$ denotes the expectation with respect to the invariant probability measure of q,

3) F is of class $\mathcal{C}^2$ and has bounded partial derivatives in x,

4) F is periodic with respect to τ with period T_0.

If $\varepsilon \to 0$ then the processes $(X^\varepsilon(t))_{t \geq 0}$ converge in distribution to the Markov diffusion process X with generator:

$$\mathcal{L} f(x) = \int_0^\infty du \, \langle \mathbb{E}\left[F(x, q(0), \cdot).\nabla\left(F(x, q(u), \cdot).\nabla f(x)\right)\right]\rangle_\tau. \qquad (4.10)$$

where $\langle \cdot \rangle_\tau$ stands for an averaging over a period in τ.

A convergence in distribution means that for any continuous bounded functional Ψ from $\mathcal{C}$ into $\mathbb{R}$ we have

$$\mathbb{E}[\Psi(X^\varepsilon)] \xrightarrow{\varepsilon \to 0} \mathbb{E}[\Psi(X)].$$

Remark 4.23. The infinitesimal generator also reads:

$$\mathcal{L} = \sum_{i,j=1}^d a_{ij}(x) \frac{\partial^2}{\partial x_i \partial x_j} + \sum_{i=1}^d b_i(x) \frac{\partial}{\partial x_i}$$

with

$$a_{ij}(x) = \int_0^\infty du \left\langle \mathbb{E}\left[F_i(x,q(0),\cdot)F_j(x,q(u),\cdot)\right]\right\rangle_\tau,$$

$$b_i(x) = \sum_{j=1}^d \int_0^\infty du \left\langle \mathbb{E}\left[F_j(x,q(0),\cdot)\frac{\partial F_i}{\partial x_j}(x,q(u),\cdot)\right]\right\rangle_\tau.$$

Remark 4.24. We can also consider the case when F depends continuously on the macroscopic time variable t in (4.9). We then get the same result with the limit process described as a time-inhomogeneous Markov process with generator $\mathcal{L}_t$ defined as above.

Remark 4.25. The periodicity condition 4) can be removed if we assume instead of 4):

4bis) Assume that the limits

$$\lim_{T\to\infty}\frac{1}{T}\int_{t_0}^{T+t_0} d\tau\,\mathbb{E}\left[F_i(x,q(0),\tau)F_j(x,q(u),\tau)\right]$$

$$\lim_{T\to\infty}\frac{1}{T}\int_{t_0}^{T+t_0} d\tau\,\mathbb{E}\left[F_j(x,q(0),\tau)\frac{\partial F_i}{\partial x_j}(x,q(u),\tau)\right]$$

exist uniformly with respect to x in a compact, are independent on t_0 and are integrable with respect to u. We then denote

$$a_{ij}(x) = \int_0^\infty du\left(\lim_{T\to\infty}\frac{1}{T}\int_0^T d\tau\,\mathbb{E}\left[F_i(x,q(0),\tau)F_j(x,q(u),\tau)\right]\right)$$

$$b_i(x) = \sum_{j=1}^d \int_0^\infty du\left(\lim_{T\to\infty}\frac{1}{T}\int_0^T d\tau\,\mathbb{E}\left[F_j(x,q(0),\tau)\frac{\partial F_i}{\partial x_j}(x,q(u),\tau)\right]\right)$$

and assume that a and b are smooth enough so that there exists a unique diffusion process with generator $\mathcal{L}$ (assume Hypotheses H for instance).

In this section we shall give a formal proof of Proposition 4.22 which contains the key points. The strategy of the complete and rigorous proof is based of the theory of martingales and is sketched out in Appendix B. It also relies on the perturbed test function method.

The process $\bar{X}^\varepsilon(\cdot) := (X^\varepsilon(\cdot), q(\cdot/\varepsilon^2))$ is Markov with generator

$$\mathcal{L}^\varepsilon = \frac{1}{\varepsilon^2}Q + \frac{1}{\varepsilon}F\left(x,q,\frac{t}{\varepsilon}\right).\nabla.$$

The Kolmogorov backward equation for this process can be written as follows:

$$\frac{\partial V^\varepsilon}{\partial t} + \mathcal{L}^\varepsilon V^\varepsilon = 0, \quad t < T. \tag{4.11}$$

and we consider final conditions at $t = T$ that do not depend on q:

$$V^\varepsilon(t = T, q, x) = f(x)$$

where f is a smooth test function. We will solve (4.11) asymptotically as $\varepsilon \to 0$ by assuming the multiple scale expansion:

$$V^\varepsilon = \sum_{n=0}^{\infty} \varepsilon^n \, V_n(t, q, x, \tau)|_{\tau = t/\varepsilon} \tag{4.12}$$

To expand (4.11) in multiple scales we must replace $\frac{\partial}{\partial t}$ by $\frac{\partial}{\partial t} + \frac{1}{\varepsilon}\frac{\partial}{\partial \tau}$. Thus Eq. (4.11) becomes

$$\frac{\partial V^\varepsilon}{\partial t} + \frac{1}{\varepsilon^2}QV^\varepsilon + \frac{1}{\varepsilon}F.\nabla V^\varepsilon + \frac{1}{\varepsilon}\frac{\partial V^\varepsilon}{\partial \tau} = 0, \quad t < T, \tag{4.13}$$

Substitution of (4.12) into (4.13) yields a hierarchy of equations:

$$QV_0 = 0 \tag{4.14}$$

$$QV_1 + F.\nabla V_0 + \frac{\partial V_0}{\partial \tau} = 0 \tag{4.15}$$

$$QV_2 + F.\nabla V_1 + \frac{\partial V_1}{\partial \tau} + \frac{\partial V_0}{\partial t} = 0 \tag{4.16}$$

Taking into account the ergodicity of q, Eq. (4.14) implies that V_0 does not depend on q. Taking the expectation of (4.15) the equation can be reduced to $\frac{\partial V_0}{\partial \tau} = 0$ which shows that V_0 does not depend on τ and V_1 should satisfy:

$$QV_1 = -F(x, q, \tau).\nabla V_0(t, x) \tag{4.17}$$

We have assumed that q is ergodic and satisfies the Fredholm alternative, so Q has an inverse on the subspace of the functions that have mean zero with respect to the invariant probability measure $\mathbb{P}$. The right-hand side of Eq. (4.17) belongs to this subspace, so we can solve this equation for V_1:

$$V_1(t, x, q, \tau) = -Q^{-1}F(x, q, \tau).\nabla V_0(t, x) \tag{4.18}$$

where $-Q^{-1} = \int_0^\infty dt e^{tQ}$. We now substitute (4.18) into (4.16) and take the expectation with respect to $\mathbb{P}$ and the averaging over a period in τ. We then see that V_0 must satisfy:

$$\frac{\partial V_0}{\partial t} + \langle \mathbb{E}[F.\nabla(-Q^{-1}F.\nabla V_0)]\rangle_\tau = 0$$

This is the solvability condition for (4.16) and it is the limit backward Kolmogorov equation for the process X^ε, which takes the form:

$$\frac{\partial V_0}{\partial t} + \mathcal{L}V_0 = 0$$

with the limit infinitesimal generator:

$$\mathcal{L} = \int_0^\infty dt \langle \mathbb{E}[F.\nabla(e^{tQ}F.\nabla)] \rangle_\tau$$

Using the probabilistic interpretation of the semigroup e^{tQ} we can express $\mathcal{L}$ as (4.10).

5 Spreading of a pulse traveling through a random medium

We are interested in the following question: how the shape of a pulse has been modified when it emerges from a randomly layered medium? This analysis takes place in the general framework, based on separation of scales, introduced by G. Papanicolaou and his co-authors (see for instance [7] for the one-dimensional case or [4] for the three-dimensional case). We consider here the problem of acoustic propagation when the incident pulse wavelength is long compared to the correlation length of the random inhomogeneities but short compared to the size of the slab.

In this framework, it has already been proved in [4] that, when the random fluctuations are weak, the O'Doherty–Anstey theory is valid, i.e. the traveling pulse retains its shape up to a low spreading; furthermore, its shape is deterministic when observed from the point of view of an observer traveling at the same random speed as the wave while it is stochastic when the observer's speed is the mean speed of the wave. Here we do not assume the fluctuations to be small. The main result of this section consists in a complete description of the asymptotic law of the emerging pulse: we prove a limit theorem which shows that the pulse spreads in a deterministic way.

For simplicity we present the proof in the one-dimensional case with no macroscopic variations of the medium and the noise only appearing in the density of the medium. We refer to [11] for the result for one-dimensional media with macroscopic variations.

We consider an acoustic wave traveling in a one-dimensional random medium located in the region $0 \le x \le L$, satisfying the linear conservation laws:

$$\begin{cases} \rho(x)\dfrac{\partial u}{\partial t}(x,t) + \dfrac{\partial p}{\partial x}(x,t) = 0 \\[2mm] \dfrac{\partial p}{\partial t}(x,t) + \kappa(x)\dfrac{\partial u}{\partial x}(x,t) = 0 \end{cases} \tag{5.1}$$

here $u(x,t)$ and $p(x,t)$ are respectively the speed and pressure of the wave, whereas $\rho(x)$ and $\kappa(x)$ are the density and bulk modulus of the medium. In our simplified model we suppose that $\kappa(x)$ is constant equal to 1 and that $\rho(x) = 1 + \eta(\frac{x}{\varepsilon^2})$ where $\eta(\frac{x}{\varepsilon^2})$ is the rapidly varying random coefficient describing the inhomogeneities. Since these coefficients are positive we suppose that $|\eta|$ is less than a constant strictly less than 1. Furthermore we assume that $\eta(x)$ is stationary, centered and mixing enough. We may think for instance that $\eta(x) = f(X_x)$ where $(X_x)_{x \ge 0}$ is the Ornstein–Uhlenbeck process and f is a smooth function from $\mathbb{R}$ to $[-\delta, \delta]$, $\delta < 1$, satisfying $\int f(y)\bar{p}(y)dy$ where $\bar{p}$ is the invariant probability density (4.8) of the Ornstein–Uhlenbeck process. u (and p) satisfies the wave equation:

$$\left(1 + \eta\left(\frac{x}{\varepsilon^2}\right)\right) u_{tt} - u_{xx} = 0.$$

In order to precise our boundary conditions we introduce the right and left going waves $A = u + p$ and $B = u - p$ which satisfy the following system of equations:

$$\frac{\partial}{\partial x}\begin{pmatrix} A \\ B \end{pmatrix} = \left(\begin{pmatrix} -1 & 0 \\ 0 & 1 \end{pmatrix} + \frac{1}{2}\eta\left(\frac{x}{\varepsilon^2}\right)\begin{pmatrix} -1 & -1 \\ 1 & 1 \end{pmatrix}\right)\frac{\partial}{\partial t}\begin{pmatrix} A \\ B \end{pmatrix}. \quad (5.2)$$

The slab of medium we are considering is located in the region $0 \le x \le L$ and at $t = 0$ an incident pulse is generated at the interface $x = 0$ between the random medium and the outside homogeneous medium. According to previous works [7] or [4] we choose a pulse which is broad compared to the size of the random inhomogeneities but short compared to the macroscopic scale of the medium. There is no wave entering the medium at $x = L$ (see Fig. 1).

$$A(0, t) = f\left(\frac{t}{\varepsilon}\right), \qquad B(L, t) = 0 \quad (5.3)$$

where f is a function whose Fourier transform $\hat{f}$ belongs to $L^1 \cap L^2$.

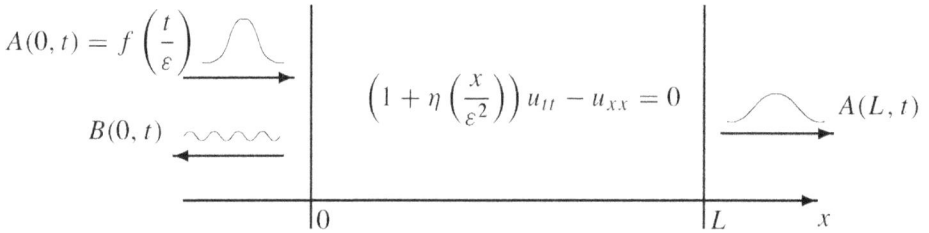

Figure 1. Spreading of a pulse

We are interested in the transmitted pulse $A(L, t)$ around the arrival time $t = L$ and in the same scale as the entering pulse $A(0, t)$; therefore the quantity of interest is the windowed signal $A(L, L + \varepsilon\sigma)_{\sigma \in (-\infty, \infty)}$ which will be given by the following centered and rescaled quantities:

$$A^{\varepsilon}(x, \sigma) = A(x, x + \varepsilon\sigma), \qquad B^{\varepsilon}(x, \sigma) = B(x, -x + \varepsilon\sigma). \quad (5.4)$$

The solution of (5.2)+(5.3) takes place in an infinite-dimensional space because of the variable t. So we perform the scaled Fourier transforms:

$$\check{A}^{\varepsilon}(x, \omega) = \int e^{i\omega\sigma} A^{\varepsilon}(x, \sigma)\, d\sigma, \qquad \check{B}^{\varepsilon}(x, \omega) = \int e^{i\omega\sigma} B^{\varepsilon}(x, \sigma)\, d\sigma.$$

In the frequency domain, with the change of variables (5.4), Eq. (5.2) becomes:

$$\frac{d}{dx}\begin{pmatrix} \check{A}^\varepsilon \\ \check{B}^\varepsilon \end{pmatrix} = \frac{i\omega}{2\varepsilon}\eta\left(\frac{x}{\varepsilon^2}\right)\begin{pmatrix} 1 & e^{-2i\omega\frac{x}{\varepsilon}} \\ -e^{2i\omega\frac{x}{\varepsilon}} & -1 \end{pmatrix}\begin{pmatrix} \check{A}^\varepsilon \\ \check{B}^\varepsilon \end{pmatrix} \tag{5.5}$$

with the boundary conditions $\check{A}^\varepsilon(0,\omega) = \hat{f}(\omega)$ and $\check{B}^\varepsilon(L,\omega) = 0$. The linearity of (5.5) enables us to replace these boundary conditions by:

$$\hat{A}^\varepsilon(0,\omega) = 1, \quad \hat{B}^\varepsilon(L,\omega) = 0 \tag{5.6}$$

and obtain the following representation for the transmitted pulse:

$$A(L, L + \varepsilon\sigma) = A^\varepsilon(L,\sigma) = \frac{1}{2\pi}\int e^{-i\omega\sigma}\,\hat{f}(\omega)\hat{A}^\varepsilon(L,\omega)\,d\omega \tag{5.7}$$

where $(\hat{A}^\varepsilon, \hat{B}^\varepsilon)$ is now the solution of problem (5.5)+(5.6).

The propagator matrix $Y^\varepsilon(x,\omega)$ is the solution of equation (5.5) with the initial condition $Y^\varepsilon(0,\omega) = \mathrm{Id}_{\mathbb{C}^2}$. The process $(\hat{A}^\varepsilon(x,\omega), \hat{B}^\varepsilon(x,\omega))$ can be deduced from $(\hat{A}^\varepsilon(0,\omega), \hat{B}^\varepsilon(0,\omega))$ through the identity:

$$\begin{pmatrix} \hat{A}^\varepsilon(x,\omega) \\ \hat{B}^\varepsilon(x,\omega) \end{pmatrix} = Y^\varepsilon(x,\omega)\begin{pmatrix} \hat{A}^\varepsilon(0,\omega) \\ \hat{B}^\varepsilon(0,\omega) \end{pmatrix}. \tag{5.8}$$

The structure of the propagator matrix can be exhibited. If $(\hat{a}^\varepsilon, \hat{b}^\varepsilon)^T$ is a solution of (5.5) with the initial condition $\hat{a}^\varepsilon(0) = 1$, $\hat{b}^\varepsilon(0) = 0$, then $((\hat{b}^\varepsilon)^*, (\hat{a}^\varepsilon)^*)^T$ is another solution linearly independent from the previous one [1] and we can thus write $Y^\varepsilon(x,\omega)$ as:

$$Y^\varepsilon(x,\omega) = \begin{pmatrix} \hat{a}^\varepsilon(x,\omega) & \hat{b}^\varepsilon(x,\omega)^* \\ \hat{b}^\varepsilon(x,\omega) & \hat{a}^\varepsilon(x,\omega)^* \end{pmatrix} \tag{5.9}$$

The trace of the matrix appearing in the linear equation (5.5) being 0, we deduce that the determinant of $Y^\varepsilon(x,\omega)$ is constant and equal to 1 which implies that $|\hat{a}^\varepsilon(x,\omega)|^2 - |\hat{b}^\varepsilon(x,\omega)|^2 = 1$ for every x. Using (5.8) and (5.9) at $x = L$ and boundary conditions (5.6) we deduce that:

$$\hat{A}^\varepsilon(L,\omega) = \frac{1}{\hat{a}^\varepsilon(L,\omega)^*}, \quad \hat{B}^\varepsilon(0,\omega) = -\frac{\hat{b}^\varepsilon(L,\omega)}{\hat{a}^\varepsilon(L,\omega)^*}. \tag{5.10}$$

In particular we have the following relation of conservation of energy:

$$|\hat{A}^\varepsilon(L,\omega)|^2 + |\hat{B}^\varepsilon(0,\omega)|^2 = 1. \tag{5.11}$$

Lemma 5.1. *The transmitted pulse* $\left((A^\varepsilon(L,\sigma))_{-\infty<\sigma<\infty}\right)_{\varepsilon>0}$ *is a tight (i.e. weakly compact) family in the space of continuous trajectories equipped with the sup norm.*

---

[1] Here the star $*$ stands for complex conjugation

Proof. We must show that, for any $\delta > 0$, there exists a compact subset K of the space of continuous bounded functions such that:

$$\sup_{\varepsilon > 0} \mathbb{P}(A^\varepsilon(L, \cdot) \in K) \geq 1 - \delta.$$

On the one hand (5.11) yields that $A^\varepsilon(L, \sigma)$ is uniformly bounded by:

$$|A^\varepsilon(L, \sigma)| \leq \frac{1}{2\pi} \int |\hat{f}(\omega)| \, d\omega.$$

On the other hand the modulus of continuity

$$M^\varepsilon(\delta) = \sup_{|\sigma_1 - \sigma_2| \leq \delta} |A^\varepsilon(L, \sigma_1) - A^\varepsilon(L, \sigma_2)|$$

is bounded by

$$M^\varepsilon(\delta) \leq \int \sup_{|\sigma_1 - \sigma_2| \leq \delta} |1 - \exp(i\omega(\sigma_1 - \sigma_2))| |\hat{f}(\omega)| \, d\omega$$

which goes to zero as δ goes to zero uniformly with respect to ε. $\square$

Moreover the finite-dimensional distributions will be characterized by the moments

$$\mathbb{E}[A^\varepsilon(L, \sigma_1)^{p_1} \ldots A^\varepsilon(L, \sigma_k)^{p_k}] \tag{5.12}$$

for every real numbers $\sigma_1 < \cdots < \sigma_k$ and every integers $p_1, \ldots, p_k$.

Let us first address the first moment. Using the representation (5.7) the expectation of $A^\varepsilon(L, \sigma)$ reads:

$$\mathbb{E}[A^\varepsilon(L, \sigma)] = \frac{1}{2\pi} \int e^{-i\omega\sigma} \hat{f}(\omega) \mathbb{E}[\hat{A}^\varepsilon(L, \omega)] \, d\omega.$$

We fix some ω and denote $X_1^\varepsilon = \text{Re}(a^\varepsilon(\cdot, \omega))$, $X_2^\varepsilon = \text{Im}(a^\varepsilon(\cdot, \omega))$, $X_3^\varepsilon = \text{Re}(b^\varepsilon(\cdot, \omega))$ and $X_4^\varepsilon = \text{Im}(b^\varepsilon(\cdot, \omega))$. The $\mathbb{R}^4$-valued process X^ε satisfies the linear differential equation

$$\frac{dX^\varepsilon(x)}{dx} = \frac{1}{\varepsilon} F_\omega\left(\eta\left(\frac{x}{\varepsilon^2}\right), \frac{x}{\varepsilon}\right) X^\varepsilon(x), \tag{5.13}$$

with the initial conditions $X_1^\varepsilon(0) = 1$ and $X_{j'}^\varepsilon(0) = 0$ if $j' = 2, 3, 4$, where

$$F_\omega(\eta, h) = \frac{\omega\eta}{2} \begin{pmatrix} 0 & -1 & \sin(2\omega h) & -\cos(2\omega h) \\ 1 & 0 & \cos(2\omega h) & \sin(2\omega h) \\ \sin(2\omega h) & \cos(2\omega h) & 0 & 1 \\ -\cos(2\omega h) & \sin(2\omega h) & -1 & 0 \end{pmatrix}.$$

Applying the approximation-diffusion Theorem 4.22, we get that X^ε converges in distribution to a Markov diffusion process X characterized by an infinitesimal generator denoted by $\mathcal{L}$:

$$\mathcal{L} = \sum_{i,j=1}^{4} a_{ij}(X)\frac{\partial^2}{\partial X_i \partial X_j} + \sum_{i=1}^{4} b_i(X)\frac{\partial}{\partial X_i}$$

whose entries b_i are all vanishing and

$$a_{11} = \frac{\alpha\omega^2}{4}\left(X_2^2 + \frac{X_3^2 + X_4^2}{2}\right),$$

$$a_{22} = \frac{\alpha\omega^2}{4}\left(X_1^2 + \frac{X_3^2 + X_4^2}{2}\right),$$

$$a_{12} = a_{21} = \frac{\alpha\omega^2}{4}(-X_1 X_2),$$

where

$$\alpha = \int_0^\infty \mathbb{E}[\eta(0)\eta(x)]\,dx.$$

The expectation $\phi(x) = \mathbb{E}[(X_1(x) - iX_2(x))^{-1}]$ satisfies the equation

$$\frac{d\phi}{dx} = \mathcal{L}\phi = -\frac{\alpha\omega^2}{2}\phi, \qquad \phi(0) = 1.$$

The solution of this ODE is: $\phi(L) = \exp(-\alpha\omega^2 L/2)$. The expectation of $\hat{A}^\varepsilon(L,\omega)$ thus converges to $\phi(L)$. By Lebesgue's theorem (remember $|\hat{A}^\varepsilon| \leq 1$) the expectation of $A^\varepsilon(L,\sigma)$ converges to:

$$\mathbb{E}[A^\varepsilon(L,\sigma)] \xrightarrow{\varepsilon \to 0} \frac{1}{2\pi}\int e^{-i\omega\sigma}\hat{f}(\omega)\exp(-\alpha\omega^2 L/2)\,d\omega.$$

Let us now consider the general moment (5.12). Using the representation (5.7) for each factor A^ε, these moments can be written as multiple integrals over $n = \sum_{j=1}^{k} p_j$ frequencies:

$$\mathbb{E}\left[A^\varepsilon(L,\sigma_1)^{p_1} \dots A^\varepsilon(L,\sigma_k)^{p_k}\right]$$

$$= \frac{1}{(2\pi)^n}\int \dots \int \prod_{\substack{1 \leq j \leq k \\ 1 \leq l \leq p_j}} \hat{f}(\omega_{j,l})e^{-i\omega_{j,l}\sigma_j}$$

$$\times \mathbb{E}\left[\prod_{\substack{1 \leq j \leq k \\ 1 \leq l \leq p_j}} \hat{A}^\varepsilon(L,\omega_{j,l})\right] \prod_{\substack{1 \leq j \leq k \\ 1 \leq l \leq p_j}} d\omega_{j,l}.$$

The dependency in ε and in the randomness only appears through the quantity

$$\mathbb{E}[\hat{A}^\varepsilon(L,\omega_1) \dots \hat{A}^\varepsilon(L,\omega_n)].$$

Our problem is now to find the limit, as ε goes to 0, of these moments for n distinct frequencies. In other words we want to study the limit in distribution of

$$(\hat{A}^\varepsilon(L, \omega_1), \ldots, \hat{A}^\varepsilon(L, \omega_n))$$

which results once again from the application of a diffusion-approximation theorem. We define the n-dimensional propagator

$$Y^\varepsilon(x, \omega_1, \omega_2, \ldots, \omega_n) = \oplus_{j=1}^N Y^\varepsilon(x, \omega_j)$$

which satisfies an equation similar to (5.5) with $Y^\varepsilon(x = 0, \omega) = \mathrm{Id}_{\mathbb{C}^{2n}}$.

In order to be allowed to apply the diffusion-approximation theorem, we have to take care to consider separately the real and imaginary parts of each coefficient $\hat{a}^\varepsilon$ and $\hat{b}^\varepsilon$, so that we actually deal with a system with $4n$ linear differential equations. Denoting $X_{4j+1}^\varepsilon = \mathrm{Re}(\hat{a}^\varepsilon(\cdot, \omega_j))$, $X_{4j+2}^\varepsilon = \mathrm{Im}(\hat{a}^\varepsilon(., \omega_j))$, $X_{4j+3}^\varepsilon = \mathrm{Re}(\hat{b}^\varepsilon(\cdot, \omega_j))$ and $X_{4j+4}^\varepsilon = \mathrm{Im}(\hat{b}^\varepsilon(\cdot, \omega_j))$, $j = 1, \ldots, n$, the $\mathbb{R}^{4n}$-valued process X^ε satisfies the linear differential equation

$$\frac{dX^\varepsilon(x)}{dx} = \frac{1}{\varepsilon} F\left(\eta\left(\frac{x}{\varepsilon^2}\right), \frac{x}{\varepsilon}\right) X^\varepsilon(x), \tag{5.14}$$

with the initial conditions $X_{4j+j'}^\varepsilon(0) = 1$ if $j' = 1$, $X_{4j+j'}^\varepsilon(0) = 0$ if $j' = 2, 3, 4$, where

$$F(\eta, h) = \oplus_{j=1}^n F_{\omega_j}(\eta, h).$$

Applying the approximation-diffusion theorem 4.22, we get that X^ε converges in distribution to a Markov diffusion process X characterized by an infinitesimal generator denoted by $\mathcal{L}$:

$$\mathcal{L} = \sum_{i,i'=1}^n \sum_{j,j'=1}^4 a_{4i+j,4i'+j'}(X) \frac{\partial^2}{\partial X_{4i+j}\partial X_{4i'+j'}}$$

$$a_{4i+1,4i+1} = \frac{\alpha\omega_i^2}{4}\left(X_{4i+2}^2 + \frac{X_{4i+3}^2 + X_{4i+4}^2}{2}\right),$$

$$a_{4i+2,4i+2} = \frac{\alpha\omega_i^2}{4}\left(X_{4i+1}^2 + \frac{X_{4i+3}^2 + X_{4i+4}^2}{2}\right),$$

$$a_{4i+1,4i+2} = a_{4i+2,4i+1} = \frac{\alpha\omega_i^2}{4}(-X_{4i+1}X_{4i+2}),$$

and for $i \neq i'$,

$$a_{4i+1,4i'+1} = \frac{\alpha \omega_i \omega_{i'}}{4} \left(X_{4i+2} X_{4i'+2} \right),$$

$$a_{4i+2,4i'+2} = \frac{\alpha \omega_i \omega_{i'}}{4} \left(X_{4i+1} X_{4i'+1} \right),$$

$$a_{4i+1,4i'+2} = a_{4i'+2,4i+1} = \frac{\alpha \omega_i \omega_{i'}}{4} \left(-X_{4i+2} X_{4i'+1} \right).$$

The quantity of interest $\mathbb{E}[\hat{A}^\varepsilon(x, \omega_1) \dots \hat{A}^\varepsilon(x, \omega_n)]$ is denoted by $\phi^\varepsilon(x)$. An application of the infinitesimal generator to the expectation

$$\mathbb{E}\left[\prod_{j=1}^{n} (X_{4j+1} - i X_{4j+2})^{-1} \right]$$

gives the following equation for $\phi(x) = \lim_{\varepsilon \to 0} \phi^\varepsilon(x)$:

$$\frac{d\phi(x)}{dx} = -\frac{2\alpha \sum_k \omega_k^2 + \alpha \sum_{k \neq l} \omega_k \omega_l}{4} \phi(x)$$

with the initial condition $\phi(0) = 1$. This linear equation has a unique solution but instead of solving it and computing explicitly our moments one can easily see that it is also satisfied by $\tilde{\phi}(x) = \mathbb{E}\left[\prod_k \tilde{A}(x, \omega_k) \right]$ where

$$\tilde{A}(x, \omega) = \exp\left(i \frac{\omega \sqrt{\alpha}}{2} B_x - \frac{\omega^2 \alpha}{4} x \right)$$

and (B_x) is a standard one-dimensional Brownian motion (B_x is a Gaussian random variable with zero-mean and variance x). Therefore $\phi(L) = \tilde{\phi}(L)$ and using (5.7) the limit in law of $A^\varepsilon(L, \sigma)$ is equal to $(2\pi)^{-1} \int e^{-i\omega\sigma} \hat{f}(\omega) \tilde{A}(L, \omega) d\omega$. Interpreting $\frac{\omega \sqrt{\alpha}}{2} B_L$ as a random phase and $\exp\left(-\frac{\omega^2 \alpha}{4} L \right)$ as the Fourier transform of the centered Gaussian density with variance $\frac{\alpha L}{2}$ denoted by $G_{\frac{\alpha L}{2}}$:

$$G_{\frac{\alpha L}{2}}(\sigma) = \frac{1}{\sqrt{\pi \alpha L}} \exp\left(-\frac{\sigma^2}{\alpha L} \right)$$

we get the main result of this section:

Proposition 5.2. *The process* $(A^\varepsilon(L, \sigma))_{\sigma \in (-\infty, \infty)}$ *converges in distribution in the space of the continuous functions to* $(\tilde{A}(L, \sigma))_{\sigma \in (-\infty, \infty)}$

$$\tilde{A}(L, \sigma) = f * G_{\frac{\alpha L}{2}} \left(\sigma - \frac{\sqrt{\alpha}}{2} B_L \right) \tag{5.15}$$

which means that the initial pulse f spreads in a deterministic way through the convolution by a Gaussian density and a random Gaussian centering appears through the Brownian motion B_L.

$\bar{A}$ is what is called by physicists the "coherent part" of the wave. The energy of $\bar{A}$ is non-random and given by:

$$\mathcal{E}_{\mathrm{coh}} = \int |f * G_{\frac{\alpha L}{2}}|^2 \, d\sigma.$$

If $f(t)$ is narrowband around a high-carrier frequency:

$$f(t) = \cos(\omega_0 t) \exp(-t^2 \delta\omega^2), \qquad \delta\omega \ll \omega_0$$

then it is found that the coherent energy decays exponentially:

$$\mathcal{E}_{\mathrm{coh}}(L) = \mathcal{E}_{\mathrm{coh}}(0) \frac{1}{\sqrt{1 + \alpha\delta\omega^2 L}} \exp\left(-\frac{\alpha\omega_0^2 L}{2(1 + \alpha\delta\omega^2 L)}\right)$$

$$\simeq \mathcal{E}_{\mathrm{coh}}(0) \exp\left(-\frac{\alpha\omega_0^2 L}{2}\right).$$

We end this section by the following remarks:

- The previous analysis has been done at L fixed. It is not difficult to generalize it to the convergence in distribution of $A^\varepsilon(L, \sigma)$ as a process in σ and L (see [11] for details). The limit is again given by (5.15) which means that the random centering of the spread pulse follows the trajectory of a Brownian motion as the pulse travels into the medium.

- In the ε-scale, the energy entering the medium at $x = 0$ is equal to $\int |f(\sigma)|^2 d\sigma$. The energy exiting the medium at $x = L$, in a coherent way around time $t = L$ in the ε-scale, is equal to $\int |f * G_{\frac{\alpha L}{2}}(\sigma)|^2 d\sigma$ which is strictly less than $\int |f(\sigma)|^2 d\sigma$. We may ask the following question: do we have a part of the missing energy exiting the medium in a coherent way somewhere else or at a different time? In other words what is the limit in distribution of $A(L, L + t_0 + \varepsilon\sigma)$ for $t_0 \neq 0$ (energy exiting at $x = L$) or $B(0, t_0 + \varepsilon\sigma)$ (energy reflected at $x = 0$). A similar analysis shows that these two processes (in σ) vanish as ε goes to 0 (see [11] for details). This means that there is no other coherent energy in the ε-scale exiting the slab $[0, L]$.

6 Scattering of a monochromatic wave by a random medium

This section is devoted to the propagation of monochromatic waves. This is a very natural approach since any wave can be described as the superposition of such elementary wavetrains by Fourier transform. Let $\hat{u}^\varepsilon(x)$ be the amplitude at $x \in \mathbb{R}$ of a monochromatic wave $u^\varepsilon(t, x) = \exp(-i\omega^\varepsilon t)\hat{u}^\varepsilon(x)$ traveling in the one-dimensional medium described in Fig. 2. The medium is homogeneous outside the slab $[0, L]$ and the wave u^ε obeys the wave equation $u^\varepsilon_{tt} - u^\varepsilon_{xx} = 0$. Accordingly $\hat{u}^\varepsilon$ satisfies

$$\hat{u}^\varepsilon_{xx} + \omega^{\varepsilon 2} \hat{u}^\varepsilon = 0$$

so that it has the form

$$\hat{u}^\varepsilon(x) = e^{i\omega^\varepsilon x} + R^\varepsilon e^{-i\omega^\varepsilon x} \qquad \text{for } x \le 0$$

and

$$\hat{u}^\varepsilon(x) = T^\varepsilon e^{i\omega^\varepsilon x} \qquad \text{for } x \ge L.$$

The complex-valued random variables R^ε and T^ε are the reflection and transmission coefficients, respectively. They depend on the particular realization of η^ε, the wavenumber ω^ε and the slab width L.

Inside the slab $[0, L]$ the perturbation is nonzero. It is the realization of a random, stationary, ergodic, and zero-mean process η^ε. The dimensionless parameter $\varepsilon > 0$ characterizes the scale of the fluctuations of the random medium as well as the wavelength of the wave. We shall assume that:

$$\eta^\varepsilon(x) = \eta\left(\frac{x}{\varepsilon^2}\right), \qquad \omega^\varepsilon = \frac{\omega}{\varepsilon}$$

which means that the correlation length of the medium $\sim \varepsilon^2$ is much smaller than the wavelength $\sim \varepsilon$ which is much smaller than the length of the medium ~ 1.

The scalar field $\hat{u}^\varepsilon$ satisfies, for $x \in [0, L]$:

$$\hat{u}^\varepsilon_{xx} + \omega^{\varepsilon 2}(1 + \eta^\varepsilon(x))\hat{u}^\varepsilon = 0. \tag{6.1}$$

The continuity of $\hat{u}^\varepsilon$ and $\hat{u}^\varepsilon_x$ at $x = 0$ and $x = L$ implies that the solution $\hat{u}^\varepsilon$ satisfies the two point boundary conditions:

$$i\omega^\varepsilon \hat{u}^\varepsilon + \hat{u}^\varepsilon_x = 2i\omega^\varepsilon \text{ at } x = 0, \qquad i\omega^\varepsilon \hat{u}^\varepsilon - \hat{u}^\varepsilon_x = 0 \text{ at } x = L. \tag{6.2}$$

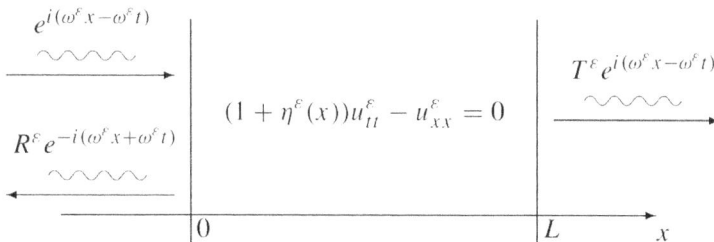

Figure 2. Scattering of a monochromatic pulse

The following statements hold true when the perturbation η is a stationary process that has finite moments of all orders and is rapidly mixing. We may think for instance that η is a Markov, stationary, ergodic process on a compact space satisfying the Fredholm alternative.

Proposition 6.1. *There exists a length* $L^\varepsilon_{\mathrm{loc}}$ *such that, with probability one,*

$$\lim_{L\to\infty} \frac{1}{L} \ln |T^\varepsilon|^2 (L, \omega) = -\frac{1}{L^\varepsilon_{\mathrm{loc}}}. \tag{6.3}$$

This length can be expanded as powers of ε:

$$\frac{1}{L^\varepsilon_{\mathrm{loc}}} = \frac{\alpha\omega^2}{2} + O(\varepsilon), \quad \alpha := \int_0^\infty du \, \mathbb{E}[\eta(0)\eta(u)]. \tag{6.4}$$

Proof. The study of the exponential behavior of the transmittivity $|T^\varepsilon|^2$ can be divided into two steps. First the localization length is shown to be equal to the inverse of the Lyapunov exponent associated to the random oscillator $v_{xx} + \omega^{\varepsilon 2}(1 + \eta^\varepsilon(x))v = 0$. Second the expansion of the Lyapunov exponent of the random oscillator is computed.

We shall first transform the boundary value problem (6.1) and (6.2) into an initial value problem. This step is similar to the analysis carried out in Section 5. Inside the perturbed slab we expand $\hat{u}$ in the form

$$\hat{u}^\varepsilon(x, \omega) = \hat{A}^\varepsilon(x, \omega)e^{i\omega^\varepsilon x} + \hat{B}^\varepsilon(x, \omega)e^{-i\omega^\varepsilon x}, \tag{6.5}$$

where $\hat{A}^\varepsilon$ and $\hat{B}^\varepsilon$ are respectively the forward (going to the right) and backward (going to the left) modes:

$$\hat{A}^\varepsilon = \frac{i\omega^\varepsilon \hat{u}^\varepsilon + \hat{u}^\varepsilon_x}{2i\omega^\varepsilon}e^{-i\omega^\varepsilon x}, \quad \hat{B}^\varepsilon = \frac{i\omega^\varepsilon \hat{u}^\varepsilon - \hat{u}^\varepsilon_x}{2i\omega^\varepsilon}e^{i\omega^\varepsilon x}.$$

The process $(\hat{A}^\varepsilon, \hat{B}^\varepsilon)$ is solution of

$$\frac{d}{dx}\begin{pmatrix} \hat{A}^\varepsilon \\ \hat{B}^\varepsilon \end{pmatrix} = P^\varepsilon(x, \omega)\begin{pmatrix} \hat{A}^\varepsilon \\ \hat{B}^\varepsilon \end{pmatrix}, \tag{6.6}$$

$$P^\varepsilon(x, \omega) = \frac{i\omega^\varepsilon}{2}\eta^\varepsilon(x)\begin{pmatrix} 1 & e^{-2i\omega^\varepsilon x} \\ -e^{2i\omega^\varepsilon x} & -1 \end{pmatrix}$$

$$= \frac{i\omega}{2\varepsilon}\eta\left(\frac{x}{\varepsilon^2}\right)\begin{pmatrix} 1 & e^{-2i\omega\frac{x}{\varepsilon}} \\ -e^{2i\omega\frac{x}{\varepsilon}} & -1 \end{pmatrix}. \tag{6.7}$$

The boundary conditions (6.2) read in terms of $\hat{A}^\varepsilon$ and $\hat{B}^\varepsilon$:

$$\hat{A}^\varepsilon(0, \omega) = 1, \quad \hat{B}^\varepsilon(L, \omega) = 0. \tag{6.8}$$

We introduce the propagator Y^ε, i.e. the fundamental matrix solution of the linear system of differential equations: $Y^\varepsilon_x = P^\varepsilon Y^\varepsilon$, $Y^\varepsilon(0) = \mathrm{Id}_{\mathbb{C}^2}$. From symmetries in

Eq. (6.6), Y^ε is of the form

$$Y^\varepsilon(x,\omega) = \begin{pmatrix} \hat{a}^\varepsilon(x,\omega) & \hat{b}^\varepsilon(x,\omega)^* \\ \hat{b}^\varepsilon(x,\omega) & \hat{a}^\varepsilon(x,\omega)^* \end{pmatrix}, \tag{6.9}$$

where $(\hat{a}^\varepsilon, \hat{b}^\varepsilon)^T$ is solution of (6.6) with the initial conditions

$$\hat{a}^\varepsilon(0,\omega) = 1, \quad \hat{b}^\varepsilon(0,\omega) = 0. \tag{6.10}$$

The modes $\hat{A}^\varepsilon$ and $\hat{B}^\varepsilon$ can be expressed in terms of the propagator:

$$\begin{pmatrix} \hat{A}^\varepsilon(x,\omega) \\ \hat{B}^\varepsilon(x,\omega) \end{pmatrix} = Y^\varepsilon(x,\omega) \begin{pmatrix} \hat{A}^\varepsilon(0,\omega) \\ \hat{B}^\varepsilon(0,\omega) \end{pmatrix}. \tag{6.11}$$

From the identity (6.11) applied for $x = L$ and the boundary conditions (6.8) we can deduce that

$$R^\varepsilon(L,\omega) = -(\hat{b}^\varepsilon/\hat{a}^\varepsilon)^*(L,\omega), \quad T^\varepsilon(L,\omega) = (1/\hat{a}^\varepsilon)^*(L,\omega). \tag{6.12}$$

The transmittivity $|T^\varepsilon|^2$ is equal to $1/|\hat{a}^\varepsilon|^2(L,\omega)$. We introduce the slow process $v^\varepsilon(x,\omega) := \hat{a}^\varepsilon(\varepsilon x,\omega)e^{i\omega^\varepsilon \varepsilon x} + \hat{b}^\varepsilon(\varepsilon x,\omega)e^{-i\omega^\varepsilon \varepsilon x}$. It satisfies the equation

$$v^\varepsilon_{xx} + \omega^2\left(1 + \eta\left(\frac{x}{\varepsilon}\right)\right)v^\varepsilon = 0$$

with the initial condition $v^\varepsilon(0) = 1$, $v^\varepsilon_x(0) = -i\omega$. Let us introduce the quantity $r^\varepsilon(x,\omega) := |v^\varepsilon|^2 + |v^\varepsilon_x|^2/\omega^2$. A straightforward calculation shows that

$$r^\varepsilon(x,\omega) = 1 + 2|\hat{a}^\varepsilon|^2(\varepsilon x,\omega).$$

By Eq. (6.12) we get a relation between r^ε and T^ε:

$$r^\varepsilon(x,\omega) = 1 + 2|T^\varepsilon(\varepsilon x,\omega)|^{-2}. \tag{6.13}$$

If $\gamma^\varepsilon(\omega)$ denotes the Lyapunov exponent that governs the exponential growth of $r^\varepsilon(x,\omega)$:

$$\gamma^\varepsilon(\omega) = \lim_{x\to\infty} \frac{1}{x} \ln r^\varepsilon(x,\omega)$$

then Eq. (6.13) insures that $|T^\varepsilon(L,\omega)|^2$ will decay as $\exp(-\gamma^\varepsilon(\omega)L/\varepsilon)$ as soon as $\gamma^\varepsilon(\omega) > 0$. In Appendix A the existence of the Lyapunov exponent $\gamma^\varepsilon(\omega)$ is proved, and its expansion with respect to ε is derived.   □

Note that α is a nonnegative real number since it is proportional to the power spectral density of the stationary random process η (Wiener–Khintchine theorem [29, p. 141]). The existence and positivity of the exponent $1/L^\varepsilon_{loc}$ can be obtained with minimal hypotheses. Kotani [25] established that a sufficient condition is that η is a stationary, ergodic process that is bounded with probability one and nondeterministic. The expansion of the localization length requires some more hypotheses about the mixing properties of η. A discussion and sufficient hypotheses are proposed in Appendix A.

Note also that the localization length of a monochromatic wave with frequency ω_0 is equal to the length that governs the exponential decay of the coherent transmitted part of a narrowband pulse with carrier frequency ω_0.

Proposition 6.2. *The square modulus of the transmission coefficient* $|T^\varepsilon(L, \omega)|^2$ *converges in distribution as a continuous process in L to the Markov process* $\tau(L, \omega)$ *whose infinitesimal generator is:*

$$\mathcal{L}_\omega = \frac{1}{2}\alpha\omega^2\tau^2(1 - \tau)\frac{\partial^2}{\partial\tau^2} - \frac{1}{2}\alpha\omega^2\tau^2\frac{\partial}{\partial\tau}. \tag{6.14}$$

Proof. The square modulus of the transmission coefficient T^ε can be expressed in terms of a random variable that is the solution of a Ricatti equation. Indeed, as a byproduct of the proof of Proposition 6.1 we find that $|T^\varepsilon|^2 = 1 - |\Gamma^\varepsilon|^2$ where $\Gamma^\varepsilon(L, \omega) = \hat{b}^\varepsilon/\hat{a}^\varepsilon(L, \omega)$ and $(\hat{a}^\varepsilon, \hat{b}^\varepsilon)$ are defined as the solutions of Eqs. (6.6) and (6.10). Differentiating $\hat{b}^\varepsilon/\hat{a}^\varepsilon$ with respect to L yields that the coefficient Γ^ε satisfies a closed-form nonlinear equation:

$$\frac{d\Gamma^\varepsilon}{dL} = -\frac{i\omega}{2\varepsilon}\eta\left(\frac{L}{\varepsilon^2}\right)\left(e^{2i\omega\frac{L}{\varepsilon}} + 2\Gamma^\varepsilon + e^{-2i\omega\frac{L}{\varepsilon}}\Gamma^{\varepsilon 2}\right), \quad \Gamma^\varepsilon(\omega, 0) = 0. \tag{6.15}$$

One then consider the process $X^\varepsilon := (r^\varepsilon, \psi^\varepsilon) := (|\Gamma^\varepsilon|^2, \arg(\Gamma^\varepsilon))$ which satisfies:

$$\frac{dX^\varepsilon}{dL}(L) = \frac{1}{\varepsilon}F\left(\eta\left(\frac{L}{\varepsilon^2}\right), X^\varepsilon(L), \frac{L}{\varepsilon}\right),$$

where F is defined by:

$$F(\eta, r, \psi, l) = \frac{\omega\eta}{2}\left(\begin{array}{c} 2\sin(\psi - 2\omega l)(r^{3/2} - r^{1/2}) \\ -2 - \cos(\psi - 2\omega l)(r^{1/2} + r^{-1/2}) \end{array}\right).$$

One then applies the diffusion-approximation Theorem 4.22 to the process $(r^\varepsilon, \psi^\varepsilon)$ which gives the result. $\qquad\qquad\qquad\square$

In particular the expectation of the square modulus of the transmission coefficient converges to $\bar{\tau}(L, \omega) := \mathbb{E}[\tau(L, \omega)]$:

$$\bar{\tau}(L, \omega) = \frac{4}{\sqrt{\pi}}\exp\left(-\frac{\alpha\omega^2 L}{8}\right)\int_0^\infty dx\frac{x^2 e^{-x^2}}{\cosh(\sqrt{\alpha\omega^2 L/2}x)}. \tag{6.16}$$

This shows that

$$\frac{1}{L}\ln\bar{\tau}(L, \omega) \overset{L\gg 1}{\simeq} -\frac{\alpha\omega^2}{8}. \tag{6.17}$$

We get actually that, for any $n \in \mathbb{N}^*$,

$$\mathbb{E}[\tau(L,\omega)^n] \overset{L \gg 1}{\simeq} \frac{c_n(\omega)}{L^{3/2}} \exp\left(-\frac{\alpha\omega^2 L}{8}\right).$$

By comparing Eq. (6.17) with Eqs. (6.3) and (6.4) we can see that the exponential behavior of the expectation of the transmittivity is different from the sample behavior of the transmittivity.

This is a quite common phenomenon when studying systems driven by random processes. We now give some heuristic arguments to complete the discussion. Let us set $\tau(L) = 2/(1 + \rho(L))$. ρ is a Markov process with infinitesimal generator:

$$\mathcal{L} = \frac{1}{2}\alpha\omega^2(\rho^2 - 1)\frac{\partial^2}{\partial\rho^2} + \alpha\omega^2\rho\frac{\partial}{\partial\rho}.$$

Applying standard tools of stochastic analysis (Itô's formula) we can represent the process ρ as the solution of the stochastic differential equation:

$$d\rho = \sqrt{\alpha}\omega\sqrt{\rho^2 - 1}\,dB_L + \alpha\omega^2\rho\,dL, \qquad \rho(0) = 1.$$

The long-range behavior is imposed by the drift so that $\rho \gg 1$ and:

$$d\rho \simeq \sqrt{\alpha}\omega\rho\,dB_L + \alpha\omega^2\rho\,dL$$

which can be solved as:

$$\rho(L) \sim \exp\left(\sqrt{\alpha}\omega B_L + \alpha\omega^2 L/2\right).$$

Please note that these identities are just heuristic! As $L \gg 1$, with probability very close to 1, we have $B_L \sim \sqrt{L}$ which is negligible compared to L, so $\rho(L) \sim \exp\left(\alpha\omega^2 L/2\right)$ and $\tau(L) \sim \exp\left(-\alpha\omega^2 L/2\right) = \exp(-L/L_{\text{loc}})$.

But, if $\sqrt{\alpha}\omega B_L < -\alpha\omega^2 L/2$, then $\rho \lesssim 1$ and $\tau \sim 1$! This is a very rare event, its probability is only $\mathbb{P}(\sqrt{\alpha}\omega B_L < -\alpha\omega^2 L/2) = \mathbb{P}(B_1 < -\sqrt{\alpha}\omega^2 L/2) \sim \exp(-\alpha\omega^2 L/8)$. But this set of rare events (= realizations of the random medium) imposes the values of the moments of the transmission coefficient.

Thus the expectation of the transmittivity is imposed by exceptional realizations of the medium. Apparently the "right" localization length is the "sample" one (6.4), in the sense that it is the one that will be observed for a typical realization of the medium. Actually we shall see that this holds true only for purely monochromatic waves.

7 Scattering of a pulse by a random medium

We consider an incoming wave from the left:

$$u_{\text{inc}}^{\varepsilon}(t, x) = \frac{1}{2\pi} \int \hat{f}^{\varepsilon}(\omega) \exp i (\omega x - \omega t) \, d\omega, \quad x \le 0, \tag{7.1}$$

where $\hat{f}^{\varepsilon} \in L^2 \cap L^1$ and contains frequencies of order ε^{-1} (i.e. wavelengths of order ε):

$$\hat{f}^{\varepsilon}(\omega) = \sqrt{\varepsilon} \hat{f}(\varepsilon \omega) \iff f^{\varepsilon}(t) = \frac{1}{\sqrt{\varepsilon}} f\left(\frac{t}{\varepsilon}\right).$$

The amplitude of the incoming pulse is normalized so that its energy is independent of ε:

$$\mathcal{E}_{\text{inc}} := \int |u_{\text{inc}}^{\varepsilon}(t, 0)|^2 dt = \frac{1}{2\pi} \int |\hat{f}^{\varepsilon}(\omega)|^2 \, d\omega = \frac{1}{2\pi} \int |\hat{f}(\omega)|^2 \, d\omega.$$

The total field in the region $x \le 0$ thus consists of the superposition of the incoming wave $u_{\text{inc}}^{\varepsilon}$ and the reflected wave:

$$u_{\text{ref}}^{\varepsilon}(t, x) = \frac{1}{2\pi \sqrt{\varepsilon}} \int \hat{f}(\omega) R^{\varepsilon}(\omega, L) \exp i \left(-\omega\frac{x}{\varepsilon} - \omega\frac{t}{\varepsilon}\right) d\omega, \quad x \le 0,$$

where $R^{\varepsilon}(\omega, L)$ is the reflection coefficient for the frequency ω/ε. The field in the region $x \ge L$ consists only of the transmitted wave that is right going:

$$u_{\text{tr}}^{\varepsilon}(t, x) = \frac{1}{2\pi \sqrt{\varepsilon}} \int \hat{f}(\omega) T^{\varepsilon}(\omega, L) \exp i \left(\omega\frac{x}{\varepsilon} - \omega\frac{t}{\varepsilon}\right) d\omega, \quad x \ge L, \tag{7.2}$$

where $T^{\varepsilon}(\omega, L)$ is the transmission coefficient for the frequency ω/ε. Inside the slab the wave has the general form:

$$u^{\varepsilon}(t, x) = \frac{1}{2\pi} \int \hat{u}^{\varepsilon}(\omega, x) \exp\left(-i\omega\frac{t}{\varepsilon}\right) d\omega, \quad 0 \le x \le L,$$

where $\hat{u}^{\varepsilon}$ satisfies the reduced wave equation:

$$\hat{u}_{xx}^{\varepsilon} + (1 + \eta^{\varepsilon}(x))\hat{u}^{\varepsilon} = 0, \quad 0 \le x \le L.$$

The total transmitted energy is:

$$\mathcal{T}^{\varepsilon}(L) = \int |u_{\text{tr}}^{\varepsilon}(t, L)|^2 \, dt = \frac{1}{2\pi} \int |\hat{f}(\omega)|^2 |T^{\varepsilon}(\omega, L)|^2 \, d\omega.$$

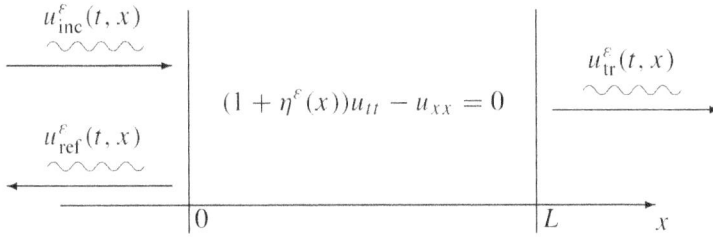

Figure 3. Scattering of a pulse

We first compute the two frequency correlation function. The following lemma is an extension of Proposition 6.2.

Lemma 7.1. *Let $\omega_1 = \omega - h\varepsilon^a/2$ and $\omega_2 = \omega + h\varepsilon^a/2$.*

1. *If $a = 0$, then the square moduli of the transmission coefficients*

$$(|T^\varepsilon(\omega_1, L)|^2, |T^\varepsilon(\omega_2, L)|^2)$$

converge in distribution to $(\tau(L, \omega - h/2), \tau(L, \omega + h/2))$ where $\tau(L, \omega - h/2)$ and $\tau(L, \omega + h/2)$ are two independent Markov processes whose infinitesimal generators are respectively $\mathcal{L}_{\omega-h/2}$ and $\mathcal{L}_{\omega+h/2}$ defined by (6.14).

2. *If $a = 1$, then the square moduli of the transmission coefficients*

$$(|T^\varepsilon(\omega_1, L)|^2, |T^\varepsilon(\omega_2, L)|^2)$$

converge in distribution to $(\tau_1(L), \tau_2(L))$ where $(\tau_1(L), \tau_2(L), \theta(L))$ is the Markov process whose infinitesimal generator is:

$$
\mathcal{L} = \frac{\alpha\omega^2}{2}\tau_1^2(1-\tau_1)\frac{\partial^2}{\partial\tau_1^2} - \frac{\alpha\omega^2}{2}\tau_1^2\frac{\partial}{\partial\tau_1} + \frac{\alpha\omega^2}{2}\tau_1^2(1-\tau_2)\frac{\partial^2}{\partial\tau_2^2} - \frac{\alpha\omega^2}{2}\tau_2^2\frac{\partial}{\partial\tau_2}
$$

$$
+ \alpha\omega^2\cos(\theta)\sqrt{(1-\tau_1)(1-\tau_2)}\,\tau_2\tau_1\frac{\partial^2}{\partial\tau_1\partial\tau_2}
$$

$$
+ 2h\frac{\partial}{\partial\theta} + \frac{\alpha\omega^2}{4}\left(\frac{(2-\tau_1)^2}{1-\tau_1} + \frac{(2-\tau_2)^2}{1-\tau_2} - 2\frac{(2-\tau_1)(2-\tau_2)}{\sqrt{(1-\tau_1)(1-\tau_2)}}\cos(\theta)\right)\frac{\partial^2}{\partial\theta^2}
$$

$$
+ \frac{\alpha\omega^2}{4}\frac{\sqrt{1-\tau_1}\tau_1(2-\tau_2)}{\sqrt{1-\tau_2}}\sin(\theta)\frac{\partial^2}{\partial\tau_1\partial\theta} \tag{7.3}
$$

$$
+ \frac{\alpha\omega^2}{4}\frac{\sqrt{1-\tau_2}\tau_2(2-\tau_1)}{\sqrt{1-\tau_1}}\sin(\theta)\frac{\partial^2}{\partial\tau_2\partial\theta}
$$

starting from $\tau_1(0) = 1$, $\tau_2(0) = 1$, and $\theta(0) = 0$.

Proof. The most interesting case is $a = 1$, since this is the correct scaling that describes the correlation of the transmission coefficients at two nearby frequencies. Let us denote $|T(\omega_j, L)|^2 = 1 - |\Gamma_j^\varepsilon|^2$ for $j = 1, 2$, where $\Gamma_j^\varepsilon(L) = \hat{b}^\varepsilon/\hat{a}^\varepsilon(\omega_j, L)$. We then introduce the four-dimensional process

$$X^\varepsilon := (r_1^\varepsilon, \psi_1^\varepsilon, r_2^\varepsilon, \psi_2^\varepsilon) := (|\Gamma_1^\varepsilon|^2, \arg(\Gamma_1^\varepsilon), |\Gamma_2^\varepsilon|^2, \arg(\Gamma_2^\varepsilon))$$

which satisfies

$$\frac{dX^\varepsilon}{dL}(L) = \frac{1}{\varepsilon} F\left(\eta\left(\frac{L}{\varepsilon^2}\right), X^\varepsilon(L), \frac{L}{\varepsilon}, L\right),$$

where F is defined by

$$F(\eta, r_1, \psi_1, r_2, \psi_2, l, L) = \frac{\omega\eta}{2}\begin{pmatrix} 2\sin(\psi_1 - 2\omega l - hL)(r_1^{3/2} - r_1^{1/2}) \\ -2 - \cos(\psi_1 - 2\omega l - hL)(r_1^{1/2} + r_1^{-1/2}) \\ 2\sin(\psi_2 - 2\omega l + hL)(r_2^{3/2} - r_2^{1/2}) \\ -2 - \cos(\psi_2 - 2\omega l + hL)(r_2^{1/2} + r_2^{-1/2}) \end{pmatrix}.$$

Applying the diffusion-approximation theorem 4.22 to the process X^ε establishes that X^ε converges to a non-homogeneous Markov process $X = (r_1, \psi_1, r_2, \psi_2)$ whose infinitesimal generator (that depends on L) can be computed explicitly. By introducing $\theta := \psi_1 - \psi_2 - 2hL$ it turns out that the process (r_1, r_2, θ) is a homogeneous Markov process whose infinitesimal generator is given by (7.3).   □

This lemma shows that the transmission coefficients corresponding to two nearby frequencies ω_1 and ω_2 are uncorrelated as soon as $|\omega_1 - \omega_2| \gg \varepsilon$. Once this result is known, it is easy to derive the asymptotic behavior of the transmittivity corresponding to the scattering of a pulse.

Proposition 7.2. *The transmittivity* $\mathcal{T}^\varepsilon(L)$ *converges in probability to* $\mathcal{T}(L)$:

$$\mathcal{T}(L) = \frac{1}{2\pi}\int |\hat{f}(\omega)|^2 \bar{\tau}(L, \omega)d\omega,$$

where $\bar{\tau}(L, \omega)$ *is the asymptotic value (6.16) of the expectation of the square modulus of the transmission coefficient* $T^\varepsilon(L, \omega)$.

Proof. Proposition 6.2 gives the limit value of the expectation of $|T^\varepsilon(L, \omega)|^2$ for one frequency ω, so that

$$\mathbb{E}\left[\mathcal{T}^\varepsilon(L)\right] \xrightarrow{\varepsilon \to 0} \frac{1}{2\pi}\int |\hat{f}(\omega)|^2 \bar{\tau}(L, \omega)d\omega.$$

Then one considers the second moment:

$$\mathbb{E}\left[\mathcal{T}^\varepsilon(L)^2\right] = \frac{1}{4\pi^2}\int\int |\hat{f}(\omega)|^2 |\hat{f}(\omega')|^2 \mathbb{E}\left[|T^\varepsilon(L, \omega)|^2 |T^\varepsilon(L, \omega')|^2\right] d\omega d\omega'.$$

The computation of this moment requires to study the two-frequency process $(|T^\varepsilon(L, \omega)|, |T^\varepsilon(L, \omega')|)$ for $\omega \neq \omega'$. Applying Lemma 7.1 one finds that $|T^\varepsilon(L, \omega)|$

and $|T^\varepsilon(L, \omega')|$ are asymptotically uncorrelated as soon as $\omega \neq \omega'$, so that

$$\mathbb{E}\left[\mathcal{T}^\varepsilon(L)^2\right] \xrightarrow{\varepsilon \to 0} \frac{1}{4\pi^2} \int\int |\hat{f}(\omega)|^2 |\hat{f}(\omega')|^2 \bar{\tau}(L, \omega)\bar{\tau}(L, \omega')$$

$$= \left(\frac{1}{2\pi} \int |\hat{f}(\omega)|^2 \bar{\tau}(L, \omega)\, d\omega\right)^2,$$

which proves the convergence of $\mathcal{T}^\varepsilon(L)$ to $\mathcal{T}(L)$ in $L^2(\mathbb{P})$:

$$\mathbb{E}\left[\left(\mathcal{T}^\varepsilon(L) - \mathcal{T}(L)\right)^2\right] = \mathbb{E}\left[\mathcal{T}^\varepsilon(L)^2\right] - 2\mathbb{E}\left[\mathcal{T}^\varepsilon(L)\right]\mathcal{T}(L) + \mathcal{T}(L)^2 \xrightarrow{\varepsilon \to 0} 0.$$

By the Chebychev inequality this implies a convergence in probability:

For any $\delta > 0$, $\mathbb{P}\left(|\mathcal{T}^\varepsilon(L) - \mathcal{T}(L)| > \delta\right) \leq \dfrac{\mathbb{E}\left[(\mathcal{T}^\varepsilon(L) - \mathcal{T}(L))^2\right]}{\delta^2} \xrightarrow{\varepsilon \to 0} 0.$ $\square$

Let us assume that the incoming wave is narrowband, that is to say that the spectral content $\hat{f}$ is concentrated around the carrier wavenumber ω_0 and has narrow bandwidth (smaller than 1, but larger than ε). Then $\mathcal{T}(L)$ decays exponentially with the width of the slab as:

$$\frac{1}{L} \ln \mathcal{T}(L) \stackrel{L \gg 1}{\simeq} -\frac{\alpha \omega_0^2}{8}$$

Note that this is the typical behavior of the *expected* value of the transmittivity of a monochromatic wave with wavenumber ω_0. In the time domain the localization process is self-averaging! This self-averaging is implied by the asymptotic decorrelation of the transmission coefficients at different frequencies. Actually $T^\varepsilon(\omega, L)$ and $T^\varepsilon(\omega', L)$ are correlated only if $|\omega - \omega'| \leq \varepsilon$.

We can now describe the energy content of the transmitted wave. It consists of a coherent part described by the O'Doherty–Anstey theory, with coherent energy that decays quickly as $\exp(-\alpha\omega_0^2 L/2)$. There is also an incoherent part in the transmitted wave, whose energy decays as $\exp(-\alpha\omega_0^2 L/8)$. Thus the incoherent wave contains most of the energy of the total transmitted wave in the regime $\alpha\omega_0^2 L \geq 1$. The spectral content of the incoherent wave is also worth studying as it contains much information [4].

Appendix A. The random harmonic oscillator

The random harmonic oscillator:

$$y_{tt} + \left(\kappa + \sigma\eta\left(\frac{t}{\varepsilon}\right)\right) y = 0 \tag{A.1}$$

with $\eta(t)$ a random process arises in many physical contexts such as solid state physics [28, 17, 31], vibrations in mechanical and electrical circuits [34, 36], and wave propagation in one-dimensional random media [23, 4]. The dimensionless parameter $\varepsilon > 0$ (resp. σ) characterizes the correlation length (resp. the amplitude) of the random fluctuations. The sample Lyapunov exponent governs the exponential growth of the modulation:

$$G := \lim_{t \to \infty} \frac{1}{t} \ln r(t), \quad r(t) = \sqrt{|y(t)|^2 + |y_t(t)|^2}. \tag{A.2}$$

Note that G could be random since η is random. So it should be relevant to study the mean and fluctuations of the Lyapunov exponent. For this purpose we shall analyze the normalized Lyapunov exponent which governs the exponential growth of the p-th moment of the modulation:

$$G_p := \lim_{t \to \infty} \frac{1}{pt} \ln \mathbb{E}\left[r(t)^p\right], \tag{A.3}$$

where $\mathbb{E}$ stands for the expectation with respect to the distribution of the process η. If the process were deterministic, then we would have $G_p = G$ for every p. But due to randomness this may not hold true since we can not invert the nonlinear power function "$| \cdot |^p$" and the linear statistical averaging "$\mathbb{E}[\cdot]$". The random matrix products theory applies to the problem (A.1). For instance let us assume that the random process η is piecewise constant over intervals $[n, n + 1)$ and take random values on the successive intervals. Under appropriate assumptions on the laws of the values taken by η, it is proved in Ref. [5, Theorem 4] that there exists an analytic function $g(p)$ such that:

$$\lim_{t \to \infty} \frac{1}{t} \ln \mathbb{E}\left[|r(t)|^p\right] = g(p), \tag{A.4}$$

$$\lim_{t \to \infty} \frac{1}{t} \ln r(t) = g'(0) \quad \text{almost surely}, \tag{A.5}$$

$$\frac{\ln r(t) - t g'(0)}{\sqrt{t}} \xrightarrow{\text{dist.}} \mathcal{N}(0, g''(0)). \tag{A.6}$$

Moreover the convergence is uniform for $r(t = 0)$ with unit modulus, and the function $p \mapsto g(p)/p$ is monotone increasing. This proves in particular that $G = g'(0)$ is non-random. In case of non-piecewise constant processes η, various versions of the above theorem exist which yield the same conclusion [18, 35, 6, 2]. Unfortunately the expression of $g(p)$ is very intricate, even for very simple random processes η. In the following sections we shall derive closed form expressions for the expansion of the sample Lyapunov exponent with respect to a small parameter.

We shall assume in the following that the driving random process $\eta(t)$ is built from a Markov process $m(t)$ by a smooth bounded function $\eta(t) = c(m(t))$. As a consequence η may be non-Markovian itself.

A.1 Small perturbations

We shall assume here that the perturbation is slow $\varepsilon = 1$, but weak $\sigma \ll 1$. There are two cases that should be distinguished: the case when $\kappa = \sigma$ and the case when $\kappa = 1$. Introducing polar coordinates $(r(t), \psi(t))$ as $y(t) = r(t)\cos(\psi(t))$ and $y_t(t) = r(t)\sin(\psi(t))$ the system (A.1) is equivalent to

$$r(t) = r_0 \exp\left(\int_0^t q(\psi(s), m(s))\, ds\right), \tag{A.7}$$

$$\psi_t(t) = h(\psi(t), m(t)), \tag{A.8}$$

with $q(\psi, m) = q_0(\psi) + \sigma q_1(\psi, m)$ and $h(\psi, m) = h_0(\psi) + \sigma h_1(\psi, m)$:

$$\text{if } \kappa = 1 \quad \begin{cases} q_0(\psi) = 0, & q_1(\psi, m) = -c(m)\sin(\psi)\cos(\psi), \\ h_0(\psi) = -1, & h_1(\psi, m) = -c(m)\cos^2(\psi) \end{cases}$$

$$\text{if } \kappa = \sigma \quad \begin{cases} q_0(\psi) = 0, & q_1(\psi, m) = -c(m)\sin(\psi)\cos(\psi), \\ h_0(\psi) = 0, & h_1(\psi, m) = -1 - c(m)\cos^2(\psi). \end{cases}$$

Let us assume that the process m is an ergodic Markov process with infinitesimal generator Q on a manifold $\mathbb{M}$ with invariant probability $\pi(dm)$. From Eq. (A.8) (ψ, m) is a Markov process on the state space $\mathbb{S}^1 \times \mathbb{M}$ where $\mathbb{S}^1$ denotes the circumference of the unit circle with infinitesimal generator: $\mathcal{L} = Q + h(\psi, m)\frac{\partial}{\partial \psi}$ and with invariant measure $\bar{p}(\psi, m)d\psi\pi(dm)$ where $\bar{p}$ can be obtained as the solution of $\mathcal{L}^*\bar{p} = 0$. According to the theorem of Crauel [12] the long-time behavior of $r(t)$ can be expressed in terms of the Lyapunov exponent G which is given by

$$G = \int_{\mathbb{S}^1 \times \mathbb{M}} q(\psi, m)\bar{p}(\psi, m)d\psi\pi(dm). \tag{A.9}$$

This result and the following ones hold true in particular under condition H1 [32] or H2 [3]:

H1 $\mathbb{M}$ is a finite set and Q is a finite-dimensional matrix which generates a continuous parameter irreducible, time-reversible Markov chain.

H2 $\mathbb{M}$ is a compact manifold. Q is a self-adjoint elliptic diffusion operator on $\mathbb{M}$ with zero an isolated, simple eigenvalue.

We have considered simple situations where $Q^* = Q$. Note that the result can be greatly generalized. For instance one can also work with the class of the ϕ-mixing processes with $\phi \in L^{1/2}$ (see [27, pp. 82–83]). The Lyapunov exponent G can be estimated in case of small noise using the technique introduced by Pinsky [32] under H1 and Arnold et al. [3] under H2. In the following we assume H2. The invariant probability measure is then simply the uniform distribution over $\mathbb{M}$.

We shall assume from now on that $\sigma \ll 1$ and we look for an expansion of G with respect to $\sigma \ll 1$. The strategy follows closely the one developed in Ref. [3]. We first divide the generator $\mathcal{L}$ into the sum $\mathcal{L} = \mathcal{L}_0 + \sigma \mathcal{L}_1$ with

$$\mathcal{L}_0 = Q + h_0(\psi)\frac{\partial}{\partial \psi}, \qquad \mathcal{L}_1 = h_1(\psi, m)\frac{\partial}{\partial \psi}.$$

As shown in [3] the probability density $\bar{p}$ can be expanded as $\bar{p} = \bar{p}_0 + \sigma \bar{p}_1 + \sigma^2 \bar{p}_2 + \cdots$ where $\bar{p}_0$, $\bar{p}_1$, and $\bar{p}_2$ satisfy $\mathcal{L}_0^* \bar{p}_0 = 0$ and $\mathcal{L}_0^* \bar{p}_1 + \mathcal{L}_1^* \bar{p}_0 = 0$, $\mathcal{L}_0^* \bar{p}_2 + \mathcal{L}_1^* \bar{p}_1 = 0, \ldots$ For once the expansion of p is known, it can be used in (A.9) to give the expansion of G at order 2 with respect to σ:

$$G = \int_{\mathbb{S}^1 \times \mathbb{M}} \big(q_0 \bar{p}_0 + \sigma (q_1 \bar{p}_0 + q_0 \bar{p}_1) + \sigma^2 (q_1 \bar{p}_1 + q_0 \bar{p}_2)\big)(\psi, m) d\psi \pi(dm) + O(\sigma^3).$$

$$(A.10)$$

If $\kappa = \sigma$. $\bar{p}_0$ satisfies $Q^* \bar{p}_0 = 0$. Thus $\bar{p}_0$ is the density of the uniform density on $\mathbb{S}^1 \times \mathbb{M}$: $\bar{p}_0 \equiv (2\pi)^{-1}$. For $\bar{p}_1$ we have the equation $Q^* \bar{p}_1 = -\mathcal{L}_1^* \bar{p}_0 = \partial_\psi (h_1 \bar{p}_0)$. Since $Q = Q^*$ this equation is of Poisson type $Q \bar{p}_1 = \partial_\psi (h_1 \bar{p}_0)$. Note that $\partial_\psi (h_1 \bar{p}_0)$ has zero-mean with respect to the invariant probability $\pi(dm)$ of Q, so the Poisson equation admits a solution $\bar{p}_1$. Let $p(0, m_0; t, m)$ be the transition probability density of the process $m(t)$. It is defined by the equation $\frac{\partial p}{\partial t} = Q^* p$, $p(0, m_0; t = 0, m) = \delta(m - m_0)$. In terms of p we can solve the equation for $\bar{p}_1$ to obtain

$$\bar{p}_1(\psi, m_0) = -\frac{1}{2\pi} \int_0^\infty dt \int_{\mathbb{M}} \partial_\psi h_1(\psi, m) p(0, m_0; t, m)\pi(dm).$$

Hence

$$G = \sigma^2 \int_{\mathbb{S}^1 \times \mathbb{M}} q_1 \bar{p}_1(\psi, m_0) d\psi \pi(dm_0) + O(\sigma^3)$$

$$= -\frac{\sigma^2}{2\pi} \int_{\mathbb{S}^1} d\psi \int_{\mathbb{M}} \pi(dm_0) \int_{\mathbb{M}} \pi(dm) \int_0^\infty dt q_1(\psi, m_0)\partial_\psi h_1(\psi, m) p(0, m_0; t, m)$$

$$+ O(\sigma^3).$$

Taking into account that the autocorrelation function reads

$$\mathbb{E}[m(0)m(t)] = \int_{\mathbb{M}} \pi(dm_0) \int_{\mathbb{M}} \pi(dm)c(m)c(m_0)p(0, m_0; t, m),$$

this can be simplified to give $G = \sigma^2 \alpha_0/4$ with

$$\alpha_0 = \int_0^\infty ds \mathbb{E}[c(m(0))c(m(s))] = \int_0^\infty ds \mathbb{E}[\eta(0)\eta(s)]. \qquad (A.11)$$

If $\kappa = 1$. Since h_0 is constant $= -1$, $\bar{p}_0$ is the uniform density on $\mathbb{S}^1 \times \mathbb{M}$: $\bar{p}_0 \equiv (2\pi)^{-1}$. Further $\bar{p}_1$ satisfies $\mathcal{L}_0^* \bar{p}_1 = -\mathcal{L}_1^* \bar{p}_0 = \partial_\psi (h_1 \bar{p}_0)$. Note that $\partial_\psi (h_1 \bar{p}_0)$ has zero-mean with respect to the invariant probability $\bar{p}_0 \pi(dm) d\psi$ of $\mathcal{L}_0^*$, so the Poisson equation admits a solution $\bar{p}_1$. In terms of the transition probability p it is given by

$$\bar{p}_1(\psi, m) = -\frac{1}{2\pi} \int_0^\infty dt \int_M \partial_\psi h_1(\psi + t) p(0, m_0; t, m) \pi(dm).$$

Substituting into (A.10) we obtain that $G = \sigma^2 \alpha_1/4 + O(\sigma^3)$, where α_1 is nonnegative and proportional to the power spectral density of the process m evaluated at 2-frequency:

$$\alpha_1 = \int_0^\infty ds \cos(2s) \mathbb{E}[c(m(0))c(m(s))] = \int_0^\infty ds \cos(2s) \mathbb{E}[\eta(0)\eta(s)]. \quad (A.12)$$

A.2 Fast perturbations

We consider the random harmonic oscillator:

$$y_{tt} + \omega^2 \left(1 + \eta\left(\frac{t}{\varepsilon}\right)\right) y = 0 \qquad (A.13)$$

Proposition A.1. *The Lyapunov exponent of the harmonic oscillator (A.13) can be expanded as powers of ε:*

$$G = \frac{\varepsilon \omega^2 \alpha_0}{4} + O(\varepsilon^2).$$

Proof. We introduce the rescaled process $\tilde{y}(t) := y(\varepsilon t)$ that satisfies

$$\tilde{y}_{tt} + \omega^2 \varepsilon^2 (1 + \eta(t)) \tilde{y} = 0.$$

Applying the above results establishes that the Lyapunov exponent of $\tilde{y}$ is $\tilde{G} = \varepsilon^2 \omega^2 \alpha_0/4 + O(\varepsilon^3)$ which gives the desired result. □

Appendix B. Diffusion-approximation

In this appendix we give the scheme for the rigorous proof of the diffusion approximation stated in Proposition 4.22. We first consider a simple case without periodic modulation.

Proposition B.1. *Let us consider the system*

$$\frac{dX^\varepsilon}{dt}(t) = \frac{1}{\varepsilon} F\left(X^\varepsilon(t), q\left(\frac{t}{\varepsilon^2}\right)\right), \qquad X^\varepsilon(0) = x_0 \in \mathbb{R}^d.$$

Assume that q is a Markov, stationary, ergodic process on a compact space with generator Q, satisfying the Fredholm alternative. F satisfies the centering condition: $\mathbb{E}[F(x, q(0))] = 0$ where $\mathbb{E}[\cdot]$ denotes the expectation with respect to the invariant probability measure of q. Instead of technical sharp conditions, assume also that F

is smooth and has bounded partial derivatives in x. Then the continuous processes $(X^\varepsilon(t))_{t \geq 0}$ converge in distribution to the Markov diffusion process X with generator

$$\mathcal{L} f(x) = \int_0^\infty du \mathbb{E} \left[F(x, q(0)).\nabla \left(F(x, q(u)).\nabla f(x) \right) \right].$$

Proof. For an extended version of the proof and sharp conditions we refer to [27, 19]. The process $\bar{X}^\varepsilon(\cdot) := (X^\varepsilon(\cdot), q(\cdot/\varepsilon^2))$ is Markov with generator

$$\mathcal{L}^\varepsilon = \frac{1}{\varepsilon^2} Q + \frac{1}{\varepsilon} F(x, q).\nabla.$$

This implies that, for any smooth function f, the process $f(\bar{X}^\varepsilon(t)) - f(\bar{X}^\varepsilon(s)) - \int_s^t \mathcal{L}^\varepsilon f(\bar{X}^\varepsilon(u)) du$ is a martingale. The proof consists in demonstrating the convergence of the corresponding martingale problems. It is based on the so-called perturbed test function method.

Step 1. *Perturbed test function method.* $\forall f \in \mathcal{C}_b^\infty$, $\forall K$ compact subset of $\mathbb{R}^d$, there exists a family f^ε such that:

$$\sup_{x \in K, q} |f^\varepsilon(x, q) - f(x)| \xrightarrow{\varepsilon \to 0} 0, \qquad \sup_{x \in K, q} |\mathcal{L}^\varepsilon f^\varepsilon(x, q) - \mathcal{L} f(x)| \xrightarrow{\varepsilon \to 0} 0. \text{ (B.1)}$$

Define $f^\varepsilon(x, q) = f(x) + \varepsilon f_1(x, q) + \varepsilon^2 f_2(x, q)$. Applying $\mathcal{L}^\varepsilon$ to f^ε, one gets

$$\mathcal{L}^\varepsilon f^\varepsilon = \frac{1}{\varepsilon} (Q f_1 + F(x, q).\nabla f(x)) + (Q f_2 + F.\nabla f_1(x, q)) + O(\varepsilon).$$

One then defines the corrections f_j as follows:

1. $f_1(x, q) = -Q^{-1} (F(x, q).\nabla f(x))$. This function is well-defined since Q has an inverse on the subspace of centered functions (Fredholm alternative). It also admits the representation

$$f_1(x, q) = \int_0^\infty du \mathbb{E}[F(x, q(u)).\nabla f(x) \mid q(0) = q].$$

2. $f_2(x, q) = -Q^{-1} (F.\nabla f_1(x, q) - \mathbb{E}[F.\nabla f_1(x, q(0))])$ is well-defined since the argument of Q^{-1} has zero-mean. It thus remains: $\mathcal{L}^\varepsilon f^\varepsilon = \mathbb{E}[F.\nabla f_1(x, q(0))] + O(\varepsilon)$ which proves (B.1).

Step 2. *Convergence of martingale problems.* One first establishes the tightness of the process X^ε in the space of the càd-làg functions equipped with the Skorohod topology by checking a standard criterion (see [27, Section 3.3]). Second one considers a subsequence $\varepsilon_p \to 0$ such that $X^{\varepsilon_p} \to X$. One takes $t_1 < \cdots < t_n < s < t$ and $h_1, \ldots, h_n \in \mathcal{C}_b^\infty$:

$$\mathbb{E}\left[\left(f^\varepsilon \left(X^\varepsilon(t), q\left(\frac{t}{\varepsilon^2}\right) \right) - f^\varepsilon(X^\varepsilon(s), q\left(\frac{s}{\varepsilon^2}\right) \right) \right.$$
$$\left. - \int_s^t \mathcal{L}^\varepsilon f^\varepsilon \left(X^\varepsilon(u), q\left(\frac{u}{\varepsilon^2}\right) \right) du \right) h_1(X^\varepsilon(t_1)) \ldots h_n(X^\varepsilon(t_n)) \right] = 0.$$

Taking the limit $\varepsilon_p \to 0$ yields

$$\mathbb{E}\left[\left(f(X(t)) - f(X(s)) - \int_s^t \mathcal{L}f(X(u))du\right)h_1(X(t_1))\ldots h_n(X(t_n))\right] = 0$$

which shows that X is solution of the martingale problem associated to $\mathcal{L}$. This problem is well-posed (in the sense that it has a unique solution), which proves the result.   □

Proposition B.2. *Let us consider the system*

$$\frac{dX^\varepsilon}{dt}(t) = \frac{1}{\varepsilon}F\left(X^\varepsilon(t), q(\frac{t}{\varepsilon^2}), \frac{t}{\varepsilon}\right), \qquad X^\varepsilon(0) = x_0 \in \mathbb{R}^d.$$

We assume the same hypotheses as in Proposition B.1. We assume also that $F(x, q, \tau)$ is periodic with respect to τ with period T_0 and F satisfies the centering condition: $\mathbb{E}[F(x, q(0), \tau)] = 0$ for all x and τ. Then the continuous processes $(X^\varepsilon(t))_{t \geq 0}$ converge in distribution to the Markov diffusion process X with generator

$$\mathcal{L}f(x) = \int_0^\infty du \, \langle \mathbb{E}[F(x, q(0), \cdot).\nabla \, (F(x, q(u), \cdot)\nabla f(x))]\rangle_\tau.$$

where $\langle \cdot \rangle_\tau$ stands for an averaging over a period in τ.

Proof. The proof is similar to the one of Proposition B.1. The key consists in building a suitable family of perturbed functions from a given test function.

$\forall f \in \mathcal{C}_b^\infty$, $\forall K$ compact subset of $\mathbb{R}^d$, there exists a family f^ε such that:

$$\sup_{x \in K, q, \tau} |f^\varepsilon(x, q, \tau) - f(x)| \xrightarrow{\varepsilon \to 0} 0, \qquad \sup_{x \in K, q, \tau} |\mathcal{L}^\varepsilon f^\varepsilon(x, q, \tau) - \mathcal{L}f(x)| \xrightarrow{\varepsilon \to 0} 0.$$

$$\text{(B.2)}$$

Let us introduce $\tau(t) := t \bmod T_0$. The process $\bar{X}^\varepsilon(\cdot) := (X^\varepsilon(\cdot), q(\cdot/\varepsilon^2), \tau(\cdot/\varepsilon))$ is Markov with generator

$$\mathcal{L}^\varepsilon = \frac{1}{\varepsilon^2}Q + \frac{1}{\varepsilon}F(x, q).\nabla + \frac{1}{\varepsilon}\frac{\partial}{\partial \tau}.$$

Let $f \in \mathcal{C}_b$. We define $f_1 = f_{11} + f_{12}$ where f_{11} is the same term as in the absence of a phase:

$$f_{11}(x, q, \tau) = -Q^{-1}(F(x, q, \tau).\nabla f(x))$$

while f_{12} does not depend on q so that $Qf_{12} = 0$:

$$f_{12}(x, \tau) = -\int_0^\tau \left(\mathbb{E}[F(x, q(0), s).\nabla f_{11}(x, q(0), s)] - \bar{f}_1(x)\right) ds$$

where $\bar{f}_1(x) = \frac{1}{T_0}\int_0^{T_0} \mathbb{E}[F(x,q(0),u).\nabla f_{11}(x,q(0),u)]du$. Note that f_{12} is uniformly bounded because of the correction $\bar{f}_1$. We finally define

$$f_2(x,q,\tau) = -Q^{-1}\left(F(x,q,\tau).\nabla f_1(x,q,\tau) + \frac{\partial f_1}{\partial \tau}(x,q,\tau)\right.$$
$$\left. - \mathbb{E}[F(x,q(0),\tau).\nabla f_1(x,q(0),\tau)] - \mathbb{E}\left[\frac{\partial f_1}{\partial \tau}(x,q(0),\tau)\right]\right).$$

Now we set

$$f^\varepsilon(x,q,\tau) = f(x) + \varepsilon f_1(x,q,\tau) + \varepsilon^2 f_2(x,q,\tau).$$

Applying the infinitesimal generator $\mathcal{L}^\varepsilon$ we get

$$\mathcal{L}^\varepsilon f^\varepsilon(x,q,\tau) = \mathbb{E}[F(x,q(0),\tau).\nabla f_1(x,q(0),\tau)] + \mathbb{E}\left[\frac{\partial f_1}{\partial \tau}(x,q(0),\tau)\right] + O(\varepsilon)$$

which simplifies into

$$\mathcal{L}^\varepsilon f^\varepsilon(x,q,\tau) = \bar{f}_1(x) + O(\varepsilon). \qquad \square$$

We refer to [19] for other multi-scaled versions of these propositions.

Bibliography

[I]   L. Brekhovskikh, *Waves in layered media*, Academic Press, 1980.
       [Deterministic scattering from inhomogeneous media]

[II]   S. N. Ethier and T. G. Kurtz, *Markov processes*, Wiley, New York 1986.
       [Markov processes and limit theorems]

[III] H. J. Kushner, *Approximation and weak convergence methods for random processes*, MIT
       Press, Cambridge 1984.
       [Diffusion-approximation and other limit theorems]

[IV]   V. I. Klyatskin, *Stochastic equations and waves in random media*, Nauka, Moscow 1980.
       [Waves in inhomogeneous media, by a physicist]

[V]   G. Papanicolaou, *Waves in one-dimensional random media*, in *Ecole d'été de Probabilités
       de Saint-Flour*, P. L. Hennequin, ed., Lecture Notes in Mathematics, Springer-Verlag,
       1988, pp. 205–275.
       [Waves in inhomogeneous media, by a mathematician]

References

[1]   P. W. Anderson, Absences of diffusion in certain random lattices, *Phys. Rev.* 109 (1958), 1492–1505.

[2]   L. Arnold, A formula connecting sample and moment stability of linear stochastic systems, *SIAM J. Appl. Math.* 44 (1984), 793–802.

[3]   L. Arnold, G. Papanicolaou, and V. Wihstutz, Asymptotic analysis of the Lyapunov exponent and rotation number of the random oscillator and applications, *SIAM J. Appl. Math.* 46 (1986), 427–450.

[4]   M. Asch, W. Kohler, G. Papanicolaou, M. Postel, and B. White, Frequency content of randomly scattered signals, *SIAM Rev.* 33 (1991), 519–625.

[5]   P. H. Baxendale and R. Z. Khaminskii, Stability index for products of random transformations, *Adv. Appl. Prob.* 30 (1998), 968–988.

[6]   P. Bougerol and J. Lacroix, *Products of random matrices with applications to Schrödinger operators*, Birkhäuser, Boston 1985.

[7]   R. Burridge, G. Papanicolaou, P. Sheng, and B. White, Probing a random medium with a pulse, *SIAM J. Appl. Math.* 49 (1989), 582–607.

[8]   R. Burridge and H. W. Chang, Multimode, one-dimensional wave propagation in a highly discontinuous medium, *Wave Motion* 11 (1989), 231–249.

[9]   R. Burridge, G. Papanicolaou, and B. White, One-dimensional wave propagation in a highly discontinuous medium, *Wave Motion* 10 (1988), 19–44.

[10] R. Carmona, Random Schrödinger operators, in *École d'été de probabilités de Saint-Flour XIV*, Lecture Notes in Math. 1180, Springer-Verlag, Berlin 1986, 1–124.

[11] J. F. Clouet and J. P. Fouque, Spreading of a pulse traveling in random media, *Ann. Appl. Probab.* 4 (1994), 1083–1097.

[12] H. Crauel, Lyapunov numbers of Markov solutions of linear stochastic systems, *Stochastic* 14 (1984), 11–28.

[13] F. Delyon, Y. Levy, and B. Souillard, Approach à la Borland to multidimensional localization, *Phys. Rev. Lett.* 55 (1985), 618–621.

[14] L. Erdös and H.-T. Yau, Linear Boltzmann equation as the weak coupling limit of a random Schrödinger equation, *Comm. Pure Appl. Math.* 53 (2000), 667–735.

[15] J.P. Fouque and E. Merzbach, A limit theorem for linear boundary value problems in random media, *Ann. Appl. Probab.* 4 (1994), 549–569.

[16] A. Friedman, *Partial differential equations of parabolic type*, Prenctice Hall, Englewood Cliffs, NJ, 1964.

[17] H. Frisch and S. P. Lloyd, Electron levels in a one-dimensional random lattice, *Phys. Rev.* 120 (1960), 1175–1189.

[18] H. Furstenberg, Noncommuting random products, *Trans. Amer. Math. Soc.* 108 (1963), 377–428.

[19] J. Garnier, A multi-scaled diffusion-approximation theorem. Applications to wave propagation in random media, *ESAIM Probability & Statistics* 1 (1997), 183–206.

[20] I. Goldsheid, S. Molchanov, and L. Pastur, A random homogeneous Schrödinger operator has a pure point spectrum, *Functional Anal. Appl.* 11 (1977), 1–10.

[21] A. Ishimaru, *Wave propagation and scattering in random media*, Academic Press, San Diego 1978.

[22] R. Z. Khaminskii, A limit theorem for solutions of differential equations with random right-hand side, *Theory Prob. Appl.* 11 (1966), 390–406.

[23] V. I. Klyatskin, *Stochastic equations and waves in random media*, Nauka, Moscow 1980.

[24] W. Kohler and G. Papanicolaou, Power statistics for wave propagation in one dimension and comparison with transport theory, *J. Math. Phys.* 14 (1973), 1733–1745; 15 (1974), 2186–2197.

[25] S. Kotani, Lyapunov indices determine absolutely continuous spectra of stationary random one-dimensional Schrödinger operators, in *Stochastic Analysis* (K. Ito, ed.), North Holland, 1984, 225–247.

[26] H. Kunita, *Stochastic flows and stochastic differential equations*, Cambridge University Press, Cambridge 1990.

[27] H. J. Kushner, *Approximation and weak convergence methods for random processes*, MIT Press, Cambridge 1984.

[28] M. Lax and J. C. Phillips, One-dimensional impurity bands, *Phys. Rev.* 110 (1958), 41–49.

[29] D. Middleton, *Introduction to statistical communication theory*, Mc Graw Hill Book Co., New York 1960.

[30] R. F. O'Doherty and N. A. Anstey, Reflections on amplitudes, *Geophysical Prospecting* 19 (1971), 430–458.

[31] L. A. Pastur, Spectra of random self-adjoint operators, *Russian Math. Surveys* 28 (1973), 3–64.

[32] M. A. Pinsky, Instability of the harmonic oscillator with small noise, *SIAM J. Appl. Math.* 46 (1986), 451–463.

[33] D. Revuz, *Markov chains*, North-Holland, Amsterdam 1975.

[34] R. L. Stratonovich, *Topics in the theory of random noise*, Gordon and Breach, New York 1963.

[35] V. N. Tutubalin, On limit theorems for products of random matrices, *Theory Prob. Appl.* 10 (1965), 15–27.

[36] N. G. Van Kampen, *Stochastic processes in physics and chemistry*, North Holland, Amsterdam 1981.

[37] A. Yu. Veretennikov, On polynomial mixing and the rate of convergence for stochastic differential and difference equations, *Theory Probab. Appl.* 44 (2000), 361–374.

[38] B. White, P. Sheng, and B. Nair, Localization and backscattering spectrum of seismic waves in stratified lithology, *Geophysics* 55 (1990), 1158–1165.

Lectures on parameter identification

Otared Kavian

Laboratoire de Mathématiques Appliquées (UMR 7641)
45, avenue des États Unis ; 78035 Versailles cedex, France
email: `kavian@math.uvsq.fr`

1 Introduction and motivation

The terms *Inverse Problems* and *Parameter Identification* can be considered synonymous, although the latter is probably less absconse to the non specialist. Both refer to situations in which a certain quantity, represented by a *parameter* or a *coefficient*, is to be determined via an indirect measurement, that is via the measurement or determination of a function, parameter or coefficient, which depends on the desired parameter in a more or less complicated way. A simple, elementary, example is when one determines the height of an inaccessible tower via the measurement of various angles and sides of appropriate triangulations (elementary Euclidian geometry establishes relations between the length of the sides and angles of a triangle). Another example, mathematically more involved, is when one knocks on a closed container, supposedly containing a liquid, in order to determine to what extent is it full or empty. In this case the frequencies of the resulting sound depend on the volume of the liquid or that of the air, and in a way or another by hearing the main frequencies of the resulting sound one is able to guess the height of the liquid. This is related to inverse spectral problems, where one wishes to recover some information about the coefficients of certain elliptic operators through information on the eigenvalues and some other measurements on the boundary of the eigenfunctions.

Another physical example in which determination of coefficients of an elliptic operator is of importance, is the following. Consider a domain Ω representing an inhomogeneous conducting body, with electric conductivity $a(x)$ at $x \in \Omega$. If one assumes that this body undergoes a known electric potential $\varphi(\sigma)$ at $\sigma \in \partial\Omega$, the boundary of Ω, then the electric potential $u(x)$ in Ω satisfies the elliptic equation:

$$\begin{cases} -\operatorname{div}(a(x)\nabla u(x)) = 0 & \text{in } \Omega \\ u(\sigma) = \varphi(\sigma) & \text{on } \partial\Omega. \end{cases} \tag{1.1}$$

In this setting the (local) intensity of the current is represented by

$$a(\sigma)\partial u(\sigma)/\partial \boldsymbol{n}(\sigma),$$

and the total electric power necessary to maintain the potential φ on the boundary $\partial\Omega$
will be

$$Q_a(\varphi) := \int_\Omega a(x)|\nabla u(x)|^2\, dx.$$

Now the question is the following: what can be deduced about the conductivity $a(x)$
of the body, if one knows these physical data? (The electric conductivity is the inverse
of the electric resistance). This question arises for instance when one wishes to obtain
some information about the material constituents of the body, without *destructive
testings*.

Note that, as a matter of fact, in the above problem the electric power $Q_a(\varphi)$ is
a redundant information, as it is entirely determined by the potential and the current
on the boundary: indeed if one multiplies equation (1.1) by u, after an integration by
parts (i.e. using Green's formula) one gets:

$$\int_\Omega a(x)|\nabla u(x)|^2\, dx = \int_{\partial\Omega} a(\sigma)\frac{\partial u(\sigma)}{\partial n(\sigma)}\varphi(\sigma)\, d\sigma.$$

Therefore we can restrict our attention to the problem in which a known potential φ
and current $a\partial u/\partial n$ on the boundary are given, and we wish to give some information
on the function a. By inspection of a few physical examples, one is convinced that
knowledge of a single set of data, that is one given pair of potential and current, in
general would not be enough information to determine $a(x)$ in Ω. Therefore, we
may give a new version of the above question, which is mathematically more precise:
can one determine uniquely the conductivity $a(x)$ if for all φ (in a certain functional
space) one knows the current $a\partial u/\partial n$, where u satisfies (1.1)? Indeed, if any such
space of φ's exists, one may also ask what is the minimal space which permits this
determination (note that since the mapping $\varphi \mapsto a\partial u/\partial n$ is linear, we may restrict
our attention to a basis $(\varphi_j)_{j\in J}$ of that space).

There exist numerous practical situations in which an identification problem may be
reduced to the determination of the coefficients of an elliptic operator. For instance the
case of a homogeneous body filling a domain Ω in which some small inhomogeneities
located in $\omega \subset \Omega$ are included, can be modeled by saying that the conductivity a
satisfies $a(x) := a_0$, a known positive constant, when x is in the homogeneous part
$\Omega \setminus \omega$ of the medium, and $a(x) := a_1$, an unknown positive constant, when $x \in \omega$.
In this case the mathematical problem would be the determination of the function a,
which amounts to the determination of both the unknown positive constant a_1 *and* the
position and shape of the subdomain ω containing the inhomogeneities.

These questions, aside being interesting in their own right, both from a mathematical point of view and in many physical applications, are of deep importance in many
other problems (inverse scattering problems, inverse spectral problems, thermography,
tomography, etc.). Later on we shall give more details in this respect.

Remark 1.1. When Ω is a bounded interval of $\mathbb{R}$, say $\Omega = (0, \ell)$, the above problem
has a quite simple answer: it is not possible to determine the conductivity $a(x)$ by

measuring potentials and currents at the endpoints of the interval. From a physical point of view this is due to the fact that different conductivities in the body, represented by the interval $(0, \ell)$, may lead to the same observed potential and current at the endpoints. The only information one may get is the average resistance, that is an equivalent homogeneous resistance one may put between the endpoints of the interval in order to obtain a given potential and current passing through the conductor. Indeed for $j = 0$ and $j = 1$ consider u_j solution of

$$-(a(x)u_j'(x))' = 0, \quad u_j(0) = 1 - j, \quad u_j(\ell) = j.$$

Then it is clear that if we denote by $M(a)$ the *harmonic mean value* of a, i.e.

$$M(a) := \left(\int_0^\ell \frac{ds}{a(s)} \right)^{-1},$$

then we have

$$u_0(x) = 1 - M(a) \int_0^x a(s)^{-1} ds, \quad u_1(x) = M(a) \int_0^x a(s)^{-1} ds.$$

Now if one solves $-(a(x)u'(x))' = 0$ with $u(0) = \alpha$ and $u(\ell) = \beta$, one has $u(x) = \alpha u_0(x) + \beta u_1(x)$, and the intensity of the current at the endpoints will be given by

$$a(0)u'(0) = a(\ell)u'(\ell) = (\beta - \alpha)M(a).$$

From this it is clear that the only information we might obtain is the value of $M(a)$, provided one applies a potential at the endpoints such that $\beta - \alpha \neq 0$. □

In the following sections we address the problem of determining the coefficient a, as well as some related questions regarding elliptic operators.

These notes have their origin in a series of lectures given at the *Universidade Federal do Rio de Janeiro* in Rio de Janeiro, Brazil and at *Université Louis Pasteur* in Strasbourg, France. I am indebted to my friends and colleagues Rolci Cipolatti, Flávio Dickstein and Nilson Costa Roberty in Rio de Janeiro, Bopeng Rao and Eric Sonnendrucker in Strasbourg, for both having me given the opportunity to give these lectures and having encouraged me to write these notes.

2 Some basic facts about linear elliptic equations

In this section we gather a few results concerning solutions of elliptic equations which should be well known before studying the related inverse problems.

In what follows we assume that $\Omega \subset \mathbb{R}^N$ is a bounded Lipschitz domain, and that $N \geq 2$; we shall denote the usual norm of $L^2(\Omega)$ by $\| \cdot \|_2$. For more details on any of the results sketched in this section the reader may refer to one of the classics on elliptic partial differential equations or D. Gilbarg and N. S. Trudinger [17], J. L. Lions [29], G. Stampacchia [37] or O. A. Ladyženskaja and N. N. Ural'ceva [27].

Thanks to the trace inequality (here we denote $\gamma_0(u) := u_{|\partial\Omega}$):

$$\exists c > 0, \ \forall u \in C^1(\overline{\Omega}) \quad \int_{\partial\Omega} |\gamma_0(u)(\sigma)|^2 \, d\sigma \le c(\|\nabla u\|_2^2 + \|u\|_2^2),$$

since $C^1(\overline{\Omega})$ is dense in $H^1(\Omega)$, the trace operator γ_0 may be extended to $H^1(\Omega)$ and by definition $H^{1/2}(\partial\Omega) := \gamma_0(H^1(\Omega))$. The Sobolev space $H^1(\Omega)$ will be equipped with the scalar product

$$(u|v)_{H^1} := \int_\Omega \nabla u(x) \cdot \nabla v(x) \, dx + \int_{\partial\Omega} u(\sigma)v(\sigma) \, d\sigma,$$

and, the associated norm $\|u\|_{H^1} := (u|u)_{H^1}^{1/2}$ is equivalent to $u \mapsto (\|\nabla u\|_2^2 + \|u\|_2^2)^{1/2}$, the usual norm of $H^1(\Omega)$.

In this setting $H_0^1(\Omega)$ is the kernel of γ_0 and the norm on $H_0^1(\Omega)$ will be $u \mapsto \|\nabla u\|_2$. The orthogonal space of $H_0^1(\Omega)$ is the space of harmonic functions on Ω:

$$H_0(\Delta) := \{v \in H^1(\Omega); \ \Delta v = 0 \text{ in } H^{-1}(\Omega)\}.$$

Since by definition $H^1(\Omega) = H_0(\Delta) \oplus H_0^1(\Omega)$, any $u \in H^1(\Omega)$ has the unique decomposition $u = \psi + u_0$ where $\psi \in H_0(\Delta)$ and $u_0 \in H_0^1(\Omega)$. This decomposition is characterized by the fact that ψ and u_0 satisfy respectively

$$\begin{cases} \Delta\psi = 0 & \text{in } \Omega \\ \psi - u = 0 & \text{on } \partial\Omega, \end{cases}$$

and

$$\begin{cases} -\Delta u_0 = -\Delta u & \text{in } \Omega \\ u_0 = 0 & \text{on } \partial\Omega. \end{cases}$$

In particular since $\|\psi\|_{H^1} \le \|u\|_{H^1}$, we have the following result:

Lemma 2.1. *For any $\varphi \in H^{1/2}(\partial\Omega)$ there exists a unique $\psi \in H^1(\Omega)$ such that*

$$\begin{cases} \Delta\psi = 0 & \text{in } \Omega \\ \psi = \varphi & \text{on } \partial\Omega. \end{cases}$$

Moreover for some positive constants c_0, c_1 depending on Ω, we have

$$c_0\|\varphi\|_{H^{1/2}(\partial\Omega)} \le \|\psi\|_{H^1} \le c_1\|\varphi\|_{H^{1/2}(\partial\Omega)}.$$

We shall denote by $\mathcal{A}_{ad}$ the set of admissible coefficients

$$\mathcal{A}_{ad} := \{a \in L^\infty(\Omega); \ \exists \varepsilon_0 > 0 \text{ such that } a \ge \varepsilon_0 \text{ a.e. on } \Omega\}. \tag{2.1}$$

For $q \in L^\infty(\Omega)$ and $a \in \mathcal{A}_{ad}$ we denote by $L_{a,q}$ the elliptic operator

$$L_{a,q}u := -\text{div}(a\nabla u) + qu, \tag{2.2}$$

which on the domain

$$D(L_{a,q}) := \left\{ u \in H_0^1(\Omega); \; -\operatorname{div}(a\nabla u) + qu \in L^2(\Omega) \right\} \qquad (2.3)$$

is a self-adjoint operator acting on $L^2(\Omega)$ (see for instance [23], chapitre 1, § 9). The operator $(L_{a,q}, D(L_{a,q}))$ has a compact resolvent, more precisely there exists $\lambda_* \in \mathbb{R}$ such that for all $\lambda > \lambda_*$, the resolvent operator

$$(\lambda I + L_{a,q})^{-1} : L^2(\Omega) \longrightarrow L^2(\Omega)$$

is compact and, as a matter of fact, the image $R((\lambda I + L_{a,q})^{-1})$ is precisely $D(L_{a,q})$.

As a consequence, $L_{a,q}$ has a nondecreasing sequence of eigenvalues $(\lambda_k)_{k \geq 1}$, each being of finite multiplicity, and eigenfunctions $(\varphi_k)_{k \geq 1}$:

$$L_{a,q}\varphi_k = \lambda_k \varphi_k, \quad \varphi_k \in D(L_{a,q}), \quad \int_\Omega \varphi_k(x)\varphi_j(x)\,dx = \delta_k^j,$$

where δ_k^j stands for the Kronecker symbol. The first eigenvalue λ_1 is simple, $\lambda_1 < \lambda_2$, and one may choose the corresponding eigenfunction φ_1 so that $\varphi_1 > 0$ in Ω. Also each eigenfunction φ_k is Lipschitz continuous in Ω, that is it belongs to $W^{1,\infty}(\Omega)$.

We shall denote by $N(L_{a,q})$ the kernel of $(L_{a,q}, D(L_{a,q}))$ and

$$H_1 := \bigoplus_{\lambda_k < 0} \mathbb{R}\varphi_k \quad \text{and} \quad H_2 := \bigoplus_{\lambda_k > 0} \mathbb{R}\varphi_k,$$

so that the space $L^2(\Omega)$ has the orthogonal decomposition

$$L^2(\Omega) = H_1 \oplus N(L_{a,q}) \oplus H_2.$$

(Note that $N(L_{a,q}) \neq \{0\}$ only if zero is an eigenvalue). Any $u \in L^2(\Omega)$ has a unique decomposition $u = u_0 + u_1 + u_2$ with $u_j \in H_j$ for $j = 1, 2$ and $u_0 \in N(L_{a,q})$, which are orthogonal projections of u on the corresponding subspaces. In particular since H_1 and $N(L_{a,q})$ are finite dimensional spaces spanned by eigenfunctions of $L_{a,q}$, one sees that $H_1 \oplus N(L_{a,q}) \subset D(L_{a,q})$ and so the projections u_0 and u_1 belong to $D(L_{a,q})$. Therefore when $u \in H_0^1(\Omega)$ the above decomposition yields that we have $u_2 \in H_0^1(\Omega)$.

Under the assumption that $N(L_{a,q}) = \{0\}$, for any given $f \in H^{-1}(\Omega)$, one can solve the problem

$$\begin{cases} -\operatorname{div}(a\nabla u) + qu = f & \text{in } H^{-1}(\Omega) \\ u = 0 & \text{on } \partial\Omega \end{cases} \qquad (2.4)$$

by a variational method. Namely, for $v \in H_0^1(\Omega)$ setting

$$J(v) := \frac{1}{2}\int_\Omega a(x)|\nabla v(x)|^2\,dx + \frac{1}{2}\int_\Omega q(x)|v(x)|^2\,dx - \int_\Omega f(x)v(x)\,dx,$$

there exists a unique $u \in H_0^1(\Omega)$ such that $u = u_1 + u_2$, with $u_j \in H_j$ and

$$J(u) = \min_{v_2 \in H_2 \cap H_0^1(\Omega)} \max_{v_1 \in H_1} J(v_1 + v_2), \tag{2.5}$$

in other words $(u_1, u_2) \in H_1 \times (H_2 \cap H_0^1(\Omega))$ is the unique saddle point of the concave-convex function $(v_1, v_2) \mapsto J(v_1 + v_2)$ (see for instance [23], chapitre 3, § 4). Moreover if we denote by $B := B_{a,q} : H^{-1}(\Omega) \longrightarrow H_0^1(\Omega)$ the mapping $f \mapsto u$, we have

$$\|u\|_{H_0^1(\Omega)} = \|Bf\|_{H_0^1(\Omega)} \le \frac{1}{\alpha_0} \|f\|_{H^{-1}(\Omega)}, \quad \text{where } \alpha_0 := \min\{|\lambda_k|;\ k \ge 1\}.$$

In the particular case of an operator $L_{a,q}$ which is coercive, that is $\lambda_1 > 0$, the functional J is strictly convex on $H_0^1(\Omega)$ and the above resolution of equation (2.4) is equivalent to using Lax–Milgram theorem (see for instance J. L. Lions [29], J. L. Lions and E. Magenes [30], K. Yosida [41]).

When zero is an eigenvalue of $(L_{a,q}, D(L_{a,q}))$, that is $N(L_{a,q}) \ne \{0\}$, since $L_{a,q}$ is selfadjoint and $R(L_{a,q}) = N(L_{a,q})^\perp = H_1 \oplus H_2$, equation (2.4) has a solution if and only if f is orthogonal to $N(L_{a,q})$ (this is a consequence of the Fredholm alternative; here $R(L_{a,q})$ denotes the range of the operator $L_{a,q}$). In this case, provided $f \in H_1 \oplus H_2 = R(L_{a,q})$, one may prove easily that (2.4) has a unique solution $u \in H_1 \oplus (H_2 \cap H_0^1(\Omega))$, and u is characterized by the equation (2.5). This means that $u \in N(L_{a,q})^\perp$ and that

$$\|u\|_{H_0^1(\Omega)} \le \frac{1}{\alpha_0} \|f\|_{H^{-1}(\Omega)}, \quad \text{where } \alpha_0 := \min\{|\lambda_k|;\ k \ge 1,\ \lambda_k \ne 0\}.$$

(However note that $u + \varphi_0$ is a solution of (2.4) for any $\varphi_0 \in N(L_{a,q})$.)

Closely related to equation (2.4) is the equation

$$\begin{cases} -\operatorname{div}(a\nabla u) + qu = 0 & \text{in } H^{-1}(\Omega) \\ u = \varphi & \text{on } \partial\Omega, \end{cases} \tag{2.6}$$

for a given $\varphi \in H^{1/2}(\partial\Omega)$. Choosing a $\psi \in H^1(\Omega)$ such that $\psi = \varphi$ on $\partial\Omega$ (for instance ψ given by Lemma 2.1), the above equation is reduced to (2.4), upon setting $u =: v + \psi$ and seeking $v \in H_0^1(\Omega)$ which solves

$$-\operatorname{div}(a\nabla v) + qv = \operatorname{div}(a\nabla\psi) - q\psi \ \text{ in } H^{-1}(\Omega), \quad v \in H_0^1(\Omega). \tag{2.7}$$

This equation has a unique solution when $N(L_{a,q}) = \{0\}$, for any given $\psi \in H^1(\Omega)$, and since $\|\psi\|_{H^1(\Omega)} \le c\|\varphi\|_{H^{1/2}(\partial\Omega)}$,

$$\|u\|_{H_0^1(\Omega)} \le \frac{1}{\alpha_0} \|\operatorname{div}(a\nabla\psi) - q\psi\|_{H^{-1}(\Omega)} \le \frac{c}{\alpha_0} \|\varphi\|_{H^{1/2}(\partial\Omega)},$$

where we use the fact that $\|q\psi\|_{H^{-1}(\Omega)} \leq c\|q\psi\|_2 \leq c\|\psi\|_{H^1(\Omega)} \leq c\|\varphi\|_{H^{1/2}(\partial\Omega)}$, and that

$$\|\operatorname{div}(a\nabla\psi)\|_{H^{-1}(\Omega)} = \sup\left\{-\int_\Omega a\nabla\psi \cdot \nabla w\, dx;\ w \in H_0^1(\Omega),\ \|w\|_{H_0^1(\Omega)} = 1\right\}$$

$$\leq \|a\|_\infty\|\psi\|_{H^1(\Omega)} \leq c\|\varphi\|_{H^{1/2}(\partial\Omega)}.$$

When $N(L_{a,q}) \neq \{0\}$, according to our above analysis, equation (2.7) has a solution if and only if the right hand side is orthogonal to the kernel $N(L_{a,q})$, that is if for all $\varphi_0 \in N(L_{a,q})$ one has:

$$\int_{\partial\Omega} a(\sigma)\frac{\partial\varphi_0(\sigma)}{\partial n(\sigma)}\, \psi(\sigma)\, d\sigma = \int_\Omega a\nabla\psi.\nabla\varphi_0\, dx + \int_\Omega q\psi\varphi_0\, dx = 0.$$

Since the kernel $N(L_{a,q})$ is finite dimensional, this means that equation (2.6) has a unique solution if and only if $\varphi \in H^{1/2}(\partial\Omega)$ is orthogonal to the finite dimensional space

$$\operatorname{span}\left\{a\frac{\partial\varphi_0}{\partial n};\ \varphi_0 \in N(L_{a,q})\right\}.$$

Actually this space and $N(L_{a,q})$ have the same dimension, at least when the coefficient a and the boundary $\partial\Omega$ are smooth enough so that, by some unique continuation theorems one may assert that the conditions $-\operatorname{div}(a\nabla w) + qw = 0$ in Ω and $w = a\partial w/\partial n = 0$ on $\partial\Omega$ imply $w \equiv 0$ (for results concerning unique continuation theorems see for instance V. Adolfsson and L. Escauriaza [2] or N. Garofalo and F. H. Lin [15]),).

In order to solve more general non homogeneous boundary value problems, that is equations such as (2.4) where the second condition is replaced with $\alpha u + \partial u/\partial n = \varphi$ on $\partial\Omega$, the following result concerning existence of a *normal trace* is useful.

Lemma 2.2. *Let $\Omega \subset \mathbb{R}^N$ be a bounded Lipschitz domain. There exists a constant $c_1 > 0$ such that if $F \in (L^2(\Omega))^N$ satisfies $\operatorname{div}(F) \in L^2(\Omega)$, then $F \cdot n$ can be defined as an element of $H^{-1/2}(\partial\Omega)$ and:*

$$\|F \cdot n\|_{H^{-1/2}(\partial\Omega)} \leq c_1(\|F\|_2 + \|\operatorname{div}(F)\|_2).$$

Proof. Assume that $F \in W^{1,\infty}(\Omega)$. Then, using an integration by parts, for any $\psi \in H^1(\Omega)$ we have

$$\int_{\partial\Omega}\psi(\sigma)F(\sigma)\cdot n(\sigma)\, d\sigma = \int_\Omega F(x)\cdot\nabla\psi(x)\, dx + \int_\Omega \operatorname{div}(F(x))\psi(x)\, dx.$$

Therefore, denoting the left hand side of the above identity as $\langle F \cdot n, \gamma_0(\psi)\rangle$, with the brackets $\langle \cdot, \cdot \rangle$ meaning the duality between $H^{-1/2}(\partial\Omega)$ and $H^{1/2}(\partial\Omega)$ and $\gamma_0 : H^1(\Omega) \longrightarrow H^{1/2}(\partial\Omega)$ being the trace operator, we have:

$$|\langle F \cdot n, \gamma_0(\psi)\rangle| \leq \|F\|_2\|\nabla\psi\|_2 + \|\operatorname{div}(F)\|_2\|\psi\|_2.$$

For a given $\varphi \in H^{1/2}(\partial\Omega)$ we may choose $\psi \in H^1(\Omega)$ as in Lemma 2.1 such that $\psi = \varphi$ on $\partial\Omega$, which satisfies $\|\psi\|_{H^1} \leq c_1 \|\varphi\|_{H^{1/2}(\partial\Omega)}$, for a constant c_1 depending only on Ω. Hence

$$|\langle F \cdot n, \varphi \rangle| \leq c_1 (\|F\|_2 + \|\mathrm{div}(F)\|_2) \|\varphi\|_{H^{1/2}(\partial\Omega)}$$

This means that

$$\|F \cdot n\|_{H^{-1/2}(\partial\Omega)} \leq c_1 (\|F\|_2 + \|\mathrm{div}(F)\|_2). \tag{2.8}$$

When F is no longer in $W^{1,\infty}(\Omega)$, using a density argument (see the classical books on Sobolev spaces, for instance R.A. Adams [1] or V.G. Maz'ja [31]) one can find a sequence $(F_k)_k$ of $W^{1,\infty}(\Omega)$ such that

$$F_k \to F \text{ in } (L^2(\Omega))^N, \quad \mathrm{div}(F_k) \to \mathrm{div}(F) \text{ in } L^2(\Omega).$$

Inequality (2.8) shows that $(F_k \cdot n)_k$ is a Cauchy sequence in $H^{-1/2}(\partial\Omega)$ whose limit, which is independent of the particular choice of the sequence $(F_k)_k$, will be denoted by $F \cdot n$. The lemma is proved. □

One of the consequences of this lemma is that whenever $u \in H^1(\Omega)$ satisfies

$$-\mathrm{div}(a\nabla u) + qu = 0 \text{ in } \Omega, \quad u = \varphi \text{ on } \Omega,$$

since $a\nabla u \in (L^2(\Omega))^N$ and $\mathrm{div}(a\nabla u) \in L^2(\Omega)$, we have

$$a\frac{\partial u}{\partial n} \in H^{-1/2}(\partial\Omega).$$

Hence we can consider the mapping

$$\varphi \mapsto a\frac{\partial u}{\partial n}$$

as a linear continuous operator from $H^{1/2}(\partial\Omega)$ into $H^{-1/2}(\partial\Omega)$. This operator, denoted by $\Lambda_{a,q}$, is the *Poincaré–Steklov operator* which we will study in the next sections.

We recall here the following local regularity result concerning solutions of elliptic boundary value problems (see for instance D. Gilbarg and N. S. Trudinger [17], G. Stampacchia [37]).

Proposition 2.3. *Let Ω be a Lipschitz domain, $\sigma \in \partial\Omega$ and $R_* > 0$ be such that $\partial\Omega \cap B(\sigma_*, R_*)$ is of class $C^{1,1}$, while $a \in A_{\mathrm{ad}} \cap W^{1,\infty}(B(\sigma_*, R_*) \cap \Omega)$. Assume that φ has its support in $B(\sigma_*, R_*) \cap \partial\Omega$ and $\varphi \in H^{3/2}(B(\sigma_*, R_*) \cap \partial\Omega)$. If $u \in H^1(\Omega)$ satisfies*

$$\begin{cases} -\mathrm{div}(a\nabla u) + qu = f & \text{in } \Omega \\ u = \varphi & \text{on } \partial\Omega \end{cases}$$

where $f \in L^2(\Omega)$, then

$$u \in H^2(\Omega \cap B(\sigma_*, R_*)), \quad a\frac{\partial u}{\partial n} \in H^{1/2}(B(\sigma_*, R_*) \cap \partial\Omega).$$

Also there exists a positive constant c_1 depending only on Ω and R_* and on the coefficients a, q, such that

$$\|u\|_{H^2(\Omega \cap B(\sigma_*, R_*))} \le c_1(\|f\|_2 + \|\varphi\|_{H^{3/2}(\Omega \cap B(\sigma_*, R_*))}).$$

This regularity result may be used to prove existence of solutions to equations such as (2.6) when φ is not anymore in $H^{1/2}(\partial\Omega)$, for instance when $\varphi \in H^{-1/2}(\partial\Omega)$. Indeed in this case one has to interpret the equation not in $H^{-1}(\Omega)$, but rather in the distributions sense, because the solution u would not be anymore in $H^1(\Omega)$ and so the term $\mathrm{div}(a\nabla u)$ would make sense only in $\mathcal{D}'(\Omega)$. For simplicity and clarity of exposition we shall consider only the case of two coefficients a, q such that $N(L_{a,q}) = \{0\}$, and we shall assume global regularity assumptions on Ω and a. However one may prove analogous local results when zero is an eigenvalue of $(L_{a,q}, D(L_{a,q}))$ or when Ω and a have only local regularity properties. In the particular case $a \equiv 1$ and $q \equiv 0$, a different proof of the following lemma may be found in J.L. Lions [29], chapitre VII, § 5.

Lemma 2.4. Let $\Omega \in \mathbb{R}^N$ be a bounded $C^{1,1}$ domain, $a \in \mathcal{A}_{\mathrm{ad}} \cap W^{1,\infty}(\Omega)$ and $q \in L^\infty(\Omega)$ be such that $N(L_{a,q}) = \{0\}$. Then for a given $\varphi \in H^{-1/2}(\partial\Omega)$, there exists a unique $u \in L^2(\Omega)$ such that $\mathrm{div}(a\nabla u) \in L^2(\Omega)$ and

$$\begin{cases} -\mathrm{div}(a\nabla u) + qu = 0 & \text{in } L^2(\Omega) \\ u = \varphi & \text{in } H^{-1/2}(\partial\Omega). \end{cases}$$

Moreover there exist two positive constants c_1, c_2 depending only on Ω and the coefficients a, q such that

$$\|\mathrm{div}(a\nabla u)\|_2 + \|u\|_2 \le c_1\|\varphi\|_{H^{-1/2}(\partial\Omega)}.$$

and $a\partial u/\partial n \in H^{-3/2}(\partial\Omega)$ satisfies

$$\left\| a\frac{\partial u}{\partial n} \right\|_{H^{-3/2}(\partial\Omega)} \le c_2\|\varphi\|_{H^{-1/2}(\partial\Omega)}.$$

Proof. Assume first that $\varphi \in H^{1/2}(\partial\Omega)$. Then there exists a unique $u \in H^1(\Omega)$ such that $u = \varphi$ on the boundary $\partial\Omega$ and

$$-\mathrm{div}(a\nabla u) + qu = 0. \tag{2.9}$$

Now let $v \in H^1_0(\Omega)$ solve

$$-\mathrm{div}(a\nabla v) + qv = u \text{ in } \Omega, \quad v = 0 \text{ on } \partial\Omega. \tag{2.10}$$

According to Proposition 2.3 we know that $v \in H^2(\Omega)$ and that

$$\left\| a\frac{\partial v}{\partial n} \right\|_{H^{1/2}(\partial\Omega)} \leq c\|v\|_{H^2(\Omega)} \leq c\|u\|_2. \tag{2.11}$$

Upon multiplying equation (2.10) by u and integrating by parts we see that since:

$$\int_\Omega (a\nabla v \cdot \nabla u + quv)\, dx = 0$$

(because $v \in H_0^1(\Omega)$ and u satisfies the homogeneous equation (2.9) in Ω), we have:

$$
\begin{aligned}
\int_\Omega u(x)^2\, dx &= -\int_{\partial\Omega} a\frac{\partial v(\sigma)}{\partial n(\sigma)} u(\sigma)\, d\sigma + \int_\Omega (a\nabla v \cdot \nabla u + quv)\, dx \\
&= -\int_{\partial\Omega} a\frac{\partial v(\sigma)}{\partial n(\sigma)} \varphi(\sigma)\, d\sigma \\
&\leq \|\varphi\|_{H^{-1/2}(\partial\Omega)} \left\| a\frac{\partial v}{\partial n} \right\|_{H^{1/2}(\partial\Omega)} \\
&\leq c\|\varphi\|_{H^{-1/2}(\partial\Omega)} \|u\|_2,
\end{aligned}
$$

where in the last step we have used the estimate (2.11). Therefore for any $\varphi \in H^{1/2}(\partial\Omega)$, if $u \in H^1(\Omega)$ solves (2.9) and $u = \varphi$ on $\partial\Omega$, we have

$$\|\mathrm{div}(a\nabla u)\|_2 + \|u\|_2 \leq c\|\varphi\|_{H^{-1/2}(\partial\Omega)}. \tag{2.12}$$

Also, in order to see that the *normal derivative* $a\partial u/\partial n$ has a meaning in $H^{-3/2}(\partial\Omega)$ we proceed as follows. Let $\psi \in H^{3/2}(\partial\Omega)$ be given and consider $w \in H^2(\Omega)$ solution to

$$\begin{cases} -\mathrm{div}(a\nabla w) + qw = 0 & \text{in } \Omega \\ w = \psi & \text{on } \partial\Omega. \end{cases} \tag{2.13}$$

Note that such a w exists in $H^1(\Omega)$ since $N(L_{a,q}) = \{0\}$. By the regularity result of Proposition 2.3 we have $w \in H^2(\Omega)$ and

$$\left\| a\frac{\partial w}{\partial n} \right\|_{H^{1/2}(\partial\Omega)} \leq c\|w\|_{H^2(\Omega)} \leq c\|\psi\|_{H^{3/2}(\partial\Omega)}.$$

Multiplying equation (2.13) by u and integrating by parts twice (or alternatively multiplying also equation (2.9) by w and then subtracting the resulted identities), one sees that

$$
\begin{aligned}
\int_{\partial\Omega} a\frac{\partial w(\sigma)}{\partial n(\sigma)} \varphi(\sigma)\, d\sigma &= \int_\Omega (a\nabla w \cdot \nabla u + qwu)\, dx \\
&= \int_{\partial\Omega} a\frac{\partial u(\sigma)}{\partial n(\sigma)} \psi(\sigma)\, d\sigma.
\end{aligned}
$$

Therefore

$$\left| \left\langle a \frac{\partial u}{\partial \boldsymbol{n}}, \psi \right\rangle \right| \leq \| \varphi \|_{H^{-1/2}(\partial\Omega)} \left\| a \frac{\partial w}{\partial \boldsymbol{n}} \right\|_{H^{1/2}(\partial\Omega)}$$

$$\leq c \| \varphi \|_{H^{-1/2}(\partial\Omega)} \| w \|_{H^{2}(\Omega)}$$

$$\leq c \| \varphi \|_{H^{-1/2}(\partial\Omega)} \| \psi \|_{H^{3/2}(\partial\Omega)}.$$

Finally from this we conclude that for any $\varphi \in H^{1/2}(\partial\Omega)$ the solution u of (2.9) satisfies

$$\left\| a \frac{\partial u}{\partial \boldsymbol{n}} \right\|_{H^{-3/2}(\partial\Omega)} \leq c \| \varphi \|_{H^{-1/2}(\partial\Omega)}. \tag{2.14}$$

Now if we assume only $\varphi \in H^{-1/2}(\partial\Omega)$, a standard regularization procedure allows us to see that the estimates (2.12) and (2.14) go through and the lemma is proved.

$\square$

3 A. P. Calderón's approach

In this first lecture we present the results of A. P. Calderón [10] which, despite being partial as far as the identification problem is concerned, contain some very interesting ideas, in particular the use of harmonic test functions of the type $u(x) := \exp(i\xi \cdot x + \eta \cdot x)$ where $\xi, \eta \in \mathbb{R}^N$ are such that $\xi \cdot \eta = 0$ and $|\xi| = |\eta|$ (see below the proof of Proposition 3.4).

We recall that by classical results on the resolution of elliptic boundary value problems (see Section 2) if $a \in \mathcal{A}_{ad}$, defined in (2.1), is given, then for each $\varphi \in H^{1/2}(\partial\Omega)$ there exists a unique $u \in H^1(\Omega)$ satisfying

$$\begin{cases} -\operatorname{div}(a(x)\nabla u(x)) = 0 & \text{in } \Omega \\ u(\sigma) = \varphi(\sigma) & \text{on } \partial\Omega. \end{cases} \tag{3.1}$$

More precisely u may be characterized by the fact that it is the unique point in the convex set

$$K_\varphi := \left\{ v \in H^1(\Omega); \ v = \varphi \text{ on } \partial\Omega \right\} \tag{3.2}$$

at which the strictly convex functional

$$J(v) := J_a(v) := \int_\Omega a(x) |\nabla v(x)|^2 \, dx \tag{3.3}$$

achieves its minimum on K_φ. We will write

$$u = \operatorname*{ArgMin}_{v \in K_\varphi} J_a(v)$$

to mean that $u \in K_\varphi$ and for all $v \in K_\varphi$ we have $J(u) \leq J(v)$. As the mapping $\varphi \mapsto u$ is linear continuous from $H^{1/2}(\partial\Omega)$ into $H^1(\Omega)$, it is a matter of simple algebra to check that $\varphi \mapsto J_a(u)$ defines a continuous quadratic form on $H^{1/2}(\partial\Omega)$ with values in $\mathbb{R}_+$. We will denote this quadratic form by

$$Q_a(\varphi) := Q(\varphi) := \int_\Omega a(x)|\nabla u(x)|^2\, dx, \quad \text{where } u \text{ satisfies (3.1)}. \qquad (3.4)$$

The *direct problem* would be: knowing the coefficient $a \in \mathcal{A}_{\text{ad}}$ study the mapping $a \mapsto Q_a(\varphi)$. It is clear that the *inverse problem*, namely trying to determine a assuming that the mapping $a \mapsto Q_a(\varphi)$ is known, would be better understood if one knows a good deal about the direct problem. In this regard our first observation is that, as one might expect, the mapping $(a, \varphi) \mapsto u$ is continuous, in fact analytic, on $\mathcal{A}_{\text{ad}} \times H^{1/2}(\partial\Omega)$.

Lemma 3.1. *For $(a, \varphi) \in \mathcal{A}_{\text{ad}} \times H^{1/2}(\partial\Omega)$ define $T(a, \varphi) := u$ where u satisfies (3.1). Then T is analytic on $\mathcal{A}_{\text{ad}} \times H^{1/2}(\partial\Omega)$.*

Proof. Let $(a_0, \varphi_0) \in \mathcal{A}_{\text{ad}} \times H^{1/2}(\partial\Omega)$ and assume that for some $\varepsilon_0 > 0$ we have $a_0 \geq \varepsilon_0$ a.e. We shall denote by $B := B_{a_0}$ the continuous operator defined for $f \in H^{-1}(\Omega)$ by $Bf := w$ where $w \in H_0^1(\Omega)$ satisfies $-\text{div}(a_0 \nabla w) = f$. Let $a = a_0 + b \in \mathcal{A}_{\text{ad}}$ where $b \in L^\infty(\Omega)$ is such that $\|b\|_\infty < \varepsilon_0$. If for $w \in H^1(\Omega)$ we set

$$A_b w := -\text{div}(b\nabla w),$$

then we have $\|A_b w\|_{H^{-1}(\Omega)} \leq \|b\|_\infty \|\nabla w\|_2$. Indeed

$$\|A_b w\|_{H^{-1}(\Omega)} = \sup\left\{ \langle A_b w, \psi \rangle;\ \psi \in H_0^1(\Omega),\ \|\psi\|_{H_0^1(\Omega)} = 1 \right\}$$

$$= \sup\left\{ \int_\Omega b(x)\nabla w(x) \cdot \nabla\psi(x)\, dx;\ \psi \in H_0^1(\Omega),\ \|\psi\|_{H_0^1(\Omega)} = 1 \right\}$$

$$\leq \|b\|_\infty \|\nabla w\|_2.$$

In particular, when $w \in H_0^1(\Omega)$, using the norm $w \mapsto \|\nabla w\|_2$ on $H_0^1(\Omega)$, we infer that BA_b is a bounded operator from $H_0^1(\Omega)$ into itself and that $\|BA_b\| \leq \|B\|\,\|b\|_\infty$. Now denote $u_0 := T(a_0, \varphi_0)$ and $u := T(a, \varphi_0)$. It is clear that $v := u - u_0$ belongs to $H_0^1(\Omega)$ and satisfies (recall that $\text{div}(a_0 \nabla u_0) = 0$)

$$-\text{div}(a_0 \nabla v) - \text{div}(b\nabla v) = \text{div}(b\nabla u_0) = -A_b u_0,$$

which, after applying B to both sides, can be written $(I + BA_b)v = -BA_b u_0$. If we assume that $\|a - a_0\|_\infty = \|b\|_\infty < \min(\|B\|^{-1}, \varepsilon_0)$ (see below to observe that $\min(\|B\|^{-1}, \varepsilon_0) = \varepsilon_0$), then $\|BA_b\| < 1$ and $(I + BA_b)$ is invertible so we can write

$$v = -(I + BA_b)^{-1} BA_b u_0 = \sum_{n=1}^\infty (-1)^n (BA_b)^n u_0,$$

and therefore

$$T(a, \varphi_0) = T(a_0, \varphi_0) + \sum_{n=1}^{\infty} (-1)^n (B A_{a-a_0})^n T(a_0, \varphi_0).$$

Remarking that $\varphi \mapsto T(a, \varphi)$ is linear, from this we conclude that T is analytic at $(a_0, \varphi_0) \in \mathcal{A}_{ad} \times H^{1/2}(\partial\Omega)$. □

Remark 3.2. We point out the following estimates concerning the dependence of u on the coefficient a. First observe that if we denote by ε_0 the essential infimum of $a_0 \in \mathcal{A}_{ad}$, that is $\varepsilon_0 := \sup \{\varepsilon > 0; \text{ meas}[a_0 < \varepsilon] = 0\}$, then $\|B_{a_0}\| \leq \varepsilon_0^{-1}$. Indeed for $f \in H^{-1}(\Omega)$ and $B_{a_0} f := w \in H_0^1(\Omega)$ satisfying $-\text{div}(a_0 \nabla w) = f$, we have

$$\varepsilon_0 \|Bf\|_{H_0^1(\Omega)}^2 \leq \int_\Omega a_0(x) |\nabla w(x)|^2 \, dx = \langle f, w \rangle \leq \|f\|_{H^{-1}(\Omega)} \|Bf\|_{H_0^1(\Omega)}.$$

In particular we note that $\|B\|^{-1} \geq \varepsilon_0$ and therefore $\min(\|B\|^{-1}, \varepsilon_0) = \varepsilon_0$. Also if $\|a - a_0\|_\infty < \varepsilon_0$, then $a \in \mathcal{A}_{ad}$ and, since $\|B A_{a-a_0}\|_{\mathcal{L}(H_0^1(\Omega))} \leq \|a - a_0\|_\infty$, the above analysis shows that

$$\|T(a, \varphi_0) - T(a_0, \varphi_0)\|_{H_0^1(\Omega)} \leq \frac{\|a - a_0\|_\infty}{\varepsilon_0 - \|a - a_0\|_\infty} \|T(a_0, \varphi_0)\|_{H_0^1(\Omega)}. \tag{3.5}$$

This inequality may be interpreted as the expression of the fact that if we set

$$\delta := \frac{\|T(a, \varphi_0) - T(a_0, \varphi_0)\|_{H_0^1(\Omega)}}{\|T(a_0, \varphi_0)\|_{H_0^1(\Omega)}} \quad \text{then} \quad \|a - a_0\|_\infty \geq \frac{\varepsilon_0 \delta}{1 + \delta}.$$

In order to obtain a *stability* result concerning the inverse problem, one should have a reverse inequality of the type $\|a - a_0\|_\infty \leq C F(\delta)$ for some continuous function F such that $F(0) = 0$. We shall see that although such results may be proved in some cases, in general their proof is much more involved: actually this difficulty is typical of inverse and parameter identification problems (see for instance G. Alessandrini [3, 5]).

Lemma 3.3. *Let* $(a_0, \varphi) \in \mathcal{A}_{ad} \times H^{1/2}(\partial\Omega)$ *and* $u_0 := T(a_0, \varphi)$. *Then for any* $b_0 \in L^\infty(\Omega)$ *we have*

$$\lim_{t \to 0^+} \frac{Q_{a_0 + tb_0}(\varphi) - Q_{a_0}(\varphi)}{t} = \int_\Omega b_0(x) |\nabla u_0(x)|^2 \, dx.$$

Proof. Setting $b := tb_0$ and $a := a_0 + tb_0$, for t small enough the assumptions in the proof of Lemma 3.1 are satisfied and using the notations used there, we have

$$Q_a(\varphi) = Q_{a_0}(\varphi) + t \int_\Omega b_0(x) |\nabla u_0(x)|^2 \, dx + 2 \int_\Omega a_0(x) \nabla u_0(x) \cdot \nabla v(x) \, dx$$

$$+ 2 \int_\Omega b(x) \nabla u_0(x) \cdot \nabla v(x) \, dx + \int_\Omega (a_0(x) + b(x)) |\nabla v(x)|^2 \, dx.$$

Now, due to the fact that $\mathrm{div}(a_0\nabla u_0) = 0$, multiplying this equation by $v \in H_0^1(\Omega)$ and integrating by parts, we remark that $\int_\Omega a_0(x)\nabla u_0(x) \cdot \nabla v(x)\,dx = 0$. On the other hand thanks to the estimate (3.5) we point out that $\|v\|_{H_0^1} \le c\|b\|_\infty = O(t)$ and therefore

$$2\int_\Omega b(x)\nabla u_0(x) \cdot \nabla v(x)\,dx + \int_\Omega (a_0(x) + b(x))|\nabla v(x)|^2\,dx = O(t^2).$$

Therefore $Q_{a_0+tb_0}(\varphi) = Q_{a_0}(\varphi) + t\int_\Omega b_0(x)|\nabla u_0(x)|^2\,dx + O(t^2)$, and the proof of the lemma is complete.   ☐

Proposition 3.4. *For $a \in \mathcal{A}_{\mathrm{ad}}$ denote by $\Phi(a) := Q_a$ the quadratic form on $H^{1/2}(\partial\Omega)$ associated to a via (3.4). Then Φ is analytic on $\mathcal{A}_{\mathrm{ad}}$. In particular for $a_0 \in \mathcal{A}_{\mathrm{ad}}$ and $b_0 \in L^\infty(\Omega)$ the derivative of Φ at a_0 is given by $\Phi'(a_0)b_0 = q$ where the quadratic form q is given by*

$$q(\varphi) := \int_\Omega b_0(x)|\nabla u_0(x)|^2\,dx,$$

where $u_0 \in H^1(\Omega)$ satisfies $\mathrm{div}(a_0\nabla u_0) = 0$ in Ω and $u_0 = \varphi$ on $\partial\Omega$. If $a_0 \in \mathcal{A}_{\mathrm{ad}}$ is constant, then Φ' is injective at a_0, that is if $b_0 \in L^\infty(\Omega)$ is such that $\Phi'(a_0)b_0 = 0$ then $b_0 \equiv 0$.

(Here the space of quadratic forms on $H^{1/2}(\partial\Omega)$ is endowed with its natural underlying norm topology: if q is a continuous quadratic form on $H^{1/2}(\partial\Omega)$ then for some unique selfadjoint operator Λ defined on $H^{1/2}(\partial\Omega)$ one has $q(\varphi) = (\Lambda\varphi|\varphi)_{H^{1/2}(\partial\Omega)}$, and the norm of q is given by $\|\Lambda\|_{\mathcal{L}(H^{1/2}(\partial\Omega))}$.)

Proof. The fact that Φ is differentiable is a consequence of the fact that by Lemma 3.3 the mapping $a_0 \mapsto Q_{a_0}$ is Gâteaux differentiable and its Gâteaux differential, namely $b \mapsto q$, is continuous. The fact that Φ is analytic follows from Lemma 3.1 and the expression of $\Phi'(a_0)$.

Now assume that a_0 is constant, so that $\mathrm{div}(a_0\nabla u) = 0$ means $\Delta u = 0$. Recall that we have denoted (see Section 2) $H_0(\Delta)$ the space of harmonic functions:

$$H_0(\Delta) := \{v \in H^1(\Omega);\ \Delta v = 0\},$$

which is actually the orthogonal space to $H_0^1(\Omega)$ in $H^1(\Omega)$. Note also that since a_0 is constant, by Lemma 3.3 we have:

$$(\Phi'(a_0)b_0)(\varphi) = \int_\Omega b_0(x)|\nabla u_0(x)|^2\,dx,$$

for $u_0 \in H_0(\Delta)$ such that $u_0 = \varphi$ on $\partial\Omega$. Therefore if $b_0 \in L^\infty(\Omega)$ is such that for all $u_0 \in H_0(\Delta)$ one has $\int_\Omega b_0(x)|\nabla u_0(x)|^2\,dx = 0$, then from this we conclude that

$$\text{for all } u_1, u_2 \in H_0(\Delta), \quad \int_\Omega b_0(x)\nabla u_1(x) \cdot \nabla u_2(x)\,dx = 0. \tag{3.6}$$

(This is the *polarization* of the quadratic form q). Now consider the special functions u_1, u_2 defined to be

$$u_1(x) := \exp((i \cdot \xi + \eta) \cdot x), \quad u_2(x) := \exp((i\xi - \eta) \cdot x), \tag{3.7}$$

where η, ξ are such that

$$\xi, \eta \in \mathbb{R}^N, \quad \xi \cdot \eta = 0, \quad |\xi| = |\eta| \neq 0. \tag{3.8}$$

(Recall that since $N \geq 2$, for any $\xi \neq 0$ there exists some $\eta \in \mathbb{R}^N$ satisfying the above conditions). One checks easily that thanks to (3.8) the functions u_j are harmonic in Ω and thus belong to $H_0(\Delta)$. However since $\nabla u_1(x) \cdot \nabla u_2(x) = -2|\xi|^2 \exp(2i\xi \cdot x)$, identity (3.6) implies that

$$\forall \xi \in \mathbb{R}^N, \quad \int_\Omega b_0(x) e^{2i\xi \cdot x} \, dx = 0 = \int_{\mathbb{R}^N} (1_\Omega(x) b_0(x)) e^{2i\xi \cdot x} \, dx.$$

This means that the Fourier transform of $1_\Omega b_0$, namely $\mathcal{F}(1_\Omega b_0)$, is zero, which in turn, using the injectivity of the Fourier transform, yields $b_0 \equiv 0$. □

Remark 3.5. If we could prove that $\Phi'(a_0)$ is a homeomorphism, then the implicit functions theorem would yield that in a neighborhood $B(a_0, \varepsilon_1)$ of a_0 the inverse problem can be solved: for $a \in B(a_0, \varepsilon_1)$ knowledge of Q_a would allow us to recover a. Unfortunately at this point we are not able to prove such a result concerning $\Phi'(a_0)$...

Remark 3.6. The above analysis contains the ingredients of an approximation process to recover a conductivity a, which is a small perturbation of a constant, when the quadratic form Q_a is known. Indeed let $a_0 \equiv 1$ (for instance) and assume that $a = 1 + b \in \mathcal{A}_{ad}$ is *close* to a_0. For $j = 1, 2$ and given $\varphi_j \in H^{1/2}(\partial\Omega)$ let $u_j, u_{0j} \in H^1(\Omega)$ satisfy

$$\text{div}(a\nabla u_j) = \text{div}(a_0 \nabla u_{0j}) = \Delta u_{0j} = 0 \text{ in } \Omega, \quad u_j = u_{0j} = \varphi_j \text{ on } \partial\Omega.$$

Since Q_a is known on $H^{1/2}(\partial\Omega)$, so is the bilinear form

$$M(\varphi_1, \varphi_2) := \frac{1}{2} [Q_a(\varphi_1 + \varphi_2) - Q_a(\varphi_1) - Q_a(\varphi_2)].$$

If we denote $v_j := u_j - u_{0j}$, as we did previously, we have $\int_\Omega \nabla u_{0j} \cdot \nabla v_k \, dx = 0$ since $v_k \in H_0^1(\Omega)$, and therefore

$$M(\varphi_1, \varphi_2) = \int_\Omega a(x) \nabla u_{01}(x) \cdot \nabla u_{02}(x) \, dx$$
$$+ \int_\Omega [(a - a_0)(\nabla u_{01} \cdot \nabla v_2 + \nabla u_{02} \cdot \nabla v_1) + a\nabla v_1 \cdot \nabla v_2] \, dx. \tag{3.9}$$

Now choose u_{0j} as in (3.7) with ξ, η satisfying (3.8) (this means that φ_j is the trace of u_{0j} on $\partial\Omega$). Then, assuming that $\Omega \subset B(0, R)$ for some $R > 0$, we obtain from (3.9)

$$2|\xi|^2 \hat{a}(2\xi) = -(2\pi)^{-N/2} M(\varphi_1, \varphi_2) + h(2\xi), \tag{3.10}$$

with

$$h(2\xi) := (2\pi)^{-N/2} \int_\Omega [(a - a_0)(\nabla u_{01} \cdot \nabla v_2 + \nabla u_{02} \cdot \nabla v_1) + a\nabla v_1 \cdot \nabla v_2] \, dx.$$

(Here we take as the definition of the Fourier transform

$$\widehat{f}(\xi) := (2\pi)^{-N/2} \int_{\mathbb{R}^N} f(x) \exp(-ix \cdot \xi) \, dx.$$

Note also that as a matter of fact we should have written $\mathcal{F}(1_\Omega a)(2\xi)$ instead of $\widehat{a}(2\xi)$, but this is not misleading in our analysis). The *error term* $h(2\xi)$ may be estimated by

$$|h(2\xi)| \le c_1 |\xi|^2 \|b\|_\infty^2 e^{2R|\xi|},$$

since $\|\nabla v_j\|_2 \le c\|b\|_\infty \|\nabla u_{0,j}\|_2 = c\|b\|_\infty |\xi|$, provided $\|b\|_\infty$ is small enough (actually if $\|b\|_\infty < 1$) to ensure that $a = a_0 + b \in \mathcal{A}_{\mathrm{ad}}$ (see the proof of Lemma 3.1). If F is such that

$$\widehat{F}(2\xi) := \frac{-(2\pi)^{-N/2} M(\varphi_1, \varphi_2)}{2|\xi|^2},$$

then we have $a(2\xi) = \widehat{F}(2\xi) + (2|\xi|^2)^{-1} h(2\xi)$ and finally

$$\widehat{a}(\xi) = \widehat{F}(\xi) + \frac{2h(\xi)}{|\xi|^2}. \tag{3.11}$$

Now consider $\rho(x) := (2\pi)^{-N/2} \exp(-|x|^2/2)$ and $\rho_\sigma(x) := \sigma^N \rho(\sigma x)$, so that convolution with ρ_σ yields an approximation of the identity when σ is large. Since $\widehat{\rho_\sigma}(x) = \widehat{\rho}(\xi/\sigma)$, and

$$\widehat{a}(\xi)\widehat{\rho}(\xi/\sigma) = \widehat{F}(\xi)\widehat{\rho}(\xi/\sigma) + \frac{2h(\xi)}{|\xi|^2}\widehat{\rho}(\xi/\sigma),$$

we have

$$a * \rho_\sigma = F * \rho_\sigma + r,$$

where the remainder r is defined via $\widehat{r}(\xi) := 2h(\xi)\widehat{\rho}(\xi/\sigma)/|\xi|^2$. Since $\|r\|_\infty \le (2\pi)^{-N/2}\|\widehat{r}\|_1$, it follows that

$$\|r\|_\infty \le c_2\|b\|_\infty^2 \int_{\mathbb{R}^N} |\widehat{\rho}(\xi/\sigma)| \exp(R|\xi|) d\xi = c_3\|b\|_\infty^2 \sigma^N \exp(\sigma^2 R^2/2).$$

Finally we see that for $\sigma > 0$ large, and $\|b\|_\infty$ small enough, then $F * \rho_\sigma$ is a *reasonable* approximation of $a * \rho_\sigma$, which in turn, since σ is large, is a *good* approximation of a.

$\square$

4 Identification on the boundary

Our second lecture is devoted to the recovery of the coefficient on the boundary. To be more specific let us introduce the following notations. For $a \in \mathcal{A}_{ad}$ (which is defined in (2.1)) and $\varphi \in H^{1/2}(\partial\Omega)$ we denote by

$$\Lambda_a(\varphi) := a \frac{\partial u}{\partial \mathbf{n}} \tag{4.1}$$

where $u \in H^1(\Omega)$ satisfies (3.1). The operator Λ_a, called the *Dirichlet to Neumann map* by some authors, and the *Poincaré–Steklov operator* by others, is a selfadjoint operator mapping $H^{1/2}(\partial\Omega) \longrightarrow H^{-1/2}(\partial\Omega)$. Indeed if $-\mathrm{div}(a\nabla u_j) = 0$ in Ω and $u_j = \varphi_j$ on the boundary for $j = 1, 2$, then multiplying by u_2 the equation satisfied by u_1, and vice versa, after integrating by parts we get

$$\langle \Lambda_a(\varphi_1), \varphi_2 \rangle = \int_\Omega a(x)\nabla u_1(x) \cdot \nabla u_2(x)\, dx$$

$$\langle \Lambda_a(\varphi_2), \varphi_1 \rangle = \int_\Omega a(x)\nabla u_2(x) \cdot \nabla u_1(x)\, dx,$$

that is $\langle \Lambda_a(\varphi_1), \varphi_2 \rangle = \langle \Lambda_a(\varphi_2), \varphi_1 \rangle$ for all $\varphi_1, \varphi_2 \in H^{1/2}(\partial\Omega)$. This implies that Λ_a is a continuous selfadjoint operator mapping $H^{1/2}(\partial\Omega)$ into $H^{-1/2}(\partial\Omega)$. Note also that Q_a being defined in (3.4), we have for all $\varphi \in H^{1/2}(\partial\Omega)$

$$Q_a(\varphi) = \langle \Lambda_a(\varphi), \varphi \rangle. \tag{4.2}$$

In this section we will show that when $a_j \in \mathcal{A}_{ad}$ is of class C^1 near the boundary $\partial\Omega$ for $j = 0, 1$, and if one has

$$\Lambda_{a_0} = \Lambda_{a_1}, \tag{4.3}$$

then $a_0 = a_1$ and $\nabla a_0 = \nabla a_1$ on the boundary. This result is due to R.V. Kohn & M. Vogelius [25] who prove actually much more: assuming that a_0 and a_1 are of class C^k near the boundary for some $k \geq 1$, then $\partial^\alpha a_0 = \partial^\alpha a_1$ on the boundary for $|\alpha| \leq k$, provided Ω has enough regularity. In particular, assuming that the boundary and the coefficients a_0, a_1 are real analytic, their result implies that $a_0 = a_1$ on $\overline{\Omega}$.

The idea of the proof is the following: assuming that (4.3) holds, but for instance $a_0(\sigma_0) > a_1(\sigma_0)$ for some $\sigma_0 \in \partial\Omega$, one considers appropriately chosen electric potentials ψ_m having an $H^{1/2}(\partial\Omega)$ norm equal to 1 and small support around σ_0, and such that $\|\psi_m\|_{H^{-1/2}} = O(m^{-1})$. Then if

$$\begin{cases} -\mathrm{div}(a_j \nabla u_{jm}) = 0 & \text{in } \Omega \\ u_{jm} = \psi_m & \text{on } \partial\Omega, \end{cases}$$

after a minute analysis of the behavior of u_{jm} as $m \to +\infty$, one shows that for large enough m's one has $Q_{a_0}(\psi_m) > Q_{a_1}(\psi_m)$.

We begin by the construction of the appropriate test functions $(\psi_m)_{m \geq 1}$.

Lemma 4.1. *Let Ω be a $C^{1,1}$ bounded domain and $\sigma_0 \in \partial\Omega$ be given. Then there exists a sequence of functions $(\psi_m)_{m\geq 1}$ such that $\psi_m \in H^{3/2}(\partial\Omega) \cap C_c^1(\partial\Omega)$ and for some positive constants c_1, c_2:*

(i) $\operatorname{supp}(\psi_{m+1}) \subset \operatorname{supp}(\psi_m)$ *and* $\bigcap_{m\geq 1} \operatorname{supp}(\psi_m) = \{\sigma_0\}$,

(ii) $\|\psi_m\|_{H^{1/2}(\partial\Omega)} = 1$,

(iii) $\dfrac{c_1}{m^{(1+2s)/2}} \leq \|\psi_m\|_{H^{-s}(\partial\Omega)} \leq \dfrac{c_2}{m^{(1+2s)/2}}$ *for* $-1 \leq s \leq 1$.

Proof. Via a local diffeomorphism we may assume that $\partial\Omega \subset \mathbb{R}^{N-1} \times \{0\}$ and that σ_0 is the origin. Consider a function $f_* \in C_c^\infty(\mathbb{R})$ such that f_* is not identically zero and

$$\operatorname{supp}(f_*) \subset [-1, +1], \qquad \int_{-1}^{+1} f_*(t)dt = 0. \tag{4.4}$$

Now for $x' := (x_1, \ldots, x_{N-1}) \in \mathbb{R}^{N-1}$ and integers $m \geq 1$ define

$$f_m(x') := \prod_{i=1}^{N-1} f_*(mx_i).$$

In the following, for a function f defined on $\mathbb{R}^{N-1}$, we denote by $\widehat{f}(\xi')$ its Fourier transform at $\xi' \in \mathbb{R}^{N-1}$.

We see that $\operatorname{supp}(f_m)$ is contained in $[-1/m, 1/m]^{N-1}$, and since $f_m(x') = f_1(mx')$, we have $\widehat{f_m}(\xi') = m^{-(N-1)}\widehat{f_1}(\xi'/m)$ and therefore for any $s \geq 0$

$$\|f_m\|_{H^{-s}(\mathbb{R}^{N-1})}^2 = \int_{\mathbb{R}^{N-1}} (1 + |\xi'|^2)^{-s}|\widehat{f_m}(\xi')|^2 d\xi'$$

$$= \frac{1}{m^{N-1}} \int_{\mathbb{R}^{N-1}} (1 + m^2|\xi'|^2)^{-s}|\widehat{f_1}(\xi')|^2 d\xi'. \tag{4.5}$$

Note that since $(1 + m^2\theta^2)^{-1} \geq m^{-2}(1 + \theta^2)^{-1}$ for $\theta \geq 0$, we have

$$\int_{\mathbb{R}^{N-1}} (1 + m^2|\xi'|^2)^{-s}|\widehat{f_1}(\xi')|^2 d\theta \geq \frac{c_3}{m^{2s}}. \tag{4.6}$$

On the other hand the mean value of f_* being zero, we have $\widehat{f_1}(0) = 0$. Whence using the fact that $\widehat{f_1}$ is analytic we have $|\cdot|^{-1}\widehat{f_1} \in \mathcal{S}(\mathbb{R}^{N-1})$ (remember that f_1 has a compact support), and recalling the inequality $(1 + m^2\theta^2)^{-1} \leq \theta^{-2}m^{-2}$ for $\theta > 0$, we obtain the upper bound for $0 \leq s \leq 1$

$$\int_{\mathbb{R}^{N-1}} (1 + m^2|\xi'|^2)^{-s}|\widehat{f_1}(\xi')|^2 d\xi' \leq \frac{1}{m^{2s}} \int_{\mathbb{R}^{N-1}} \frac{|\widehat{f_1}(\xi')|^2 d\xi'}{|\xi'|^{2s}} = \frac{c_4}{m^{2s}}. \tag{4.7}$$

This and (4.6) plugged into (4.5) yield that for $0 \leq s \leq 1$

$$c_5(s) m^{\frac{-(N-1)}{2}-s} \leq \|f_m\|_{H^{-s}(\mathbb{R}^{N-1})} \leq c_6(s) m^{\frac{-(N-1)}{2}-s}. \tag{4.8}$$

The same line of arguments shows that for any $s \geq 0$ we have

$$c_1(s)m^{\frac{-(N-1)}{2}+s} \leq \|f_m\|_{H^s(\mathbb{R}^{N-1})} \leq c_2(s)m^{\frac{-(N-1)}{2}+s}. \tag{4.9}$$

Finally from (4.9) and (4.8) we obtain for $s \geq -1$:

$$c_7(s)m^{\frac{-(N-1)}{2}+s} \leq \|f_m\|_{H^s(\mathbb{R}^{N-1})} \leq c_8(s)m^{\frac{-(N-1)}{2}+s},$$

and upon setting

$$\psi_m(x') := \frac{f_m(x')}{\|f_m\|_{H^{1/2}(\mathbb{R}^{N-1})}},$$

we conclude that for $-1 \leq s \leq 1$ we have

$$c_9(s)m^{-(1-2s)/2} \leq \|\psi_m\|_{H^s(\mathbb{R}^{N-1})} \leq c_{10}(s)m^{-(1-2s)/2},$$

and that the sequence $(\psi_m)_{m\geq 1}$ satisfies the conditions of the lemma, at least when the boundary $\partial\Omega$ is a portion of $\mathbb{R}^{N-1} \times \{0\}$. The general case is treated via a local diffeomorphism Φ and the resulting sequence $(\psi_m)_{m\geq 1}$ will inherit the regularity of Φ that is, with our assumptions, each ψ_m will be in $C^{1,1}(\partial\Omega) \cap H^{3/2}(\partial\Omega)$. □

In what follows, $\sigma_0 \in \partial\Omega$ and $(\psi_m)_{m\geq 1}$ being as in Lemma 4.1 and $a \in L^\infty(\Omega)$ so that $a \geq \varepsilon_0$ for some $\varepsilon_0 > 0$, we shall denote by $u_m \in H^1(\Omega)$ the solution of

$$\begin{cases} -\mathrm{div}(a\nabla u_m) = 0 & \text{in } \Omega \\ u_m = \psi_m & \text{on } \partial\Omega. \end{cases} \tag{4.10}$$

Our purpose in the following lemma is to show that, in some sense, u_m concentrates around σ_0, more precisely that, far from σ_0, the H^1-norm of u_m is small.

Lemma 4.2. *Assume that Ω is a bounded $C^{1,1}$ domain and that the coefficient a in (4.10) is in $W^{1,\infty}(B(\sigma_0, R) \cap \Omega)$. Then if $\omega := \Omega \setminus \overline{B}(\sigma_0, R)$, there exist $m_0 \geq 1$ and a constant $c > 0$ such that for all $m \geq m_0$ one has*

$$\|u_m\|_{H^1(\omega)} \leq \frac{c}{m}. \tag{4.11}$$

Proof. First we may take $m_0 \geq 1$ large enough so that the support of ψ_m is contained in $B(\sigma_0, R/8) \cap \partial\Omega$ for all $m \geq m_0$. Then we set $\omega_0 := \Omega \setminus \overline{B}(\sigma_0, R/4)$ and choose a cut-off function $\zeta \in C_c^\infty(\mathbb{R}^N)$ such that $0 \leq \zeta \leq 1$, and

$$\zeta \equiv 1 \text{ in } \omega, \quad \zeta \equiv 0 \text{ in } B(\sigma_0, R/2).$$

Now consider $v_m := \zeta u_m$ and observe that $v_m = 0$ on $\partial\omega_0$ and, since $\mathrm{div}(a\nabla u_m) = 0$,

$$-\mathrm{div}(a\nabla v_m) = -a\nabla\zeta \cdot \nabla u_m - \mathrm{div}(au_m\nabla\zeta).$$

Multiplying this equation by $v_m = \zeta u_m$, and integrating by parts over ω_0, we get

$$\int_{\omega_0} a(x)|\nabla v_m(x)|^2\, dx = -\int_{\omega_0} a(\nabla\zeta \cdot \nabla u_m)u_m \zeta\, dx + \int_{\omega_0} au_m \nabla\zeta \cdot \nabla(\zeta u_m)\, dx$$

$$= \int_{\omega_0} a(x)u_m(x)^2|\nabla\zeta(x)|^2\, dx,$$

and finally, observing that $u_m = v_m$ on $\omega \subset\subset \omega_0$,

$$\int_{\omega} |\nabla u_m(x)|^2\, dx \le c \int_{\Omega} |u_m(x)|^2 |\nabla\zeta(x)|^2\, dx. \tag{4.12}$$

Now we need to estimate the right hand side of (4.12) in terms of $\|\psi_m\|_{H^{-1/2}}$. To this end consider $\Psi \in H_0^1(\Omega)$ solution to

$$-\operatorname{div}(a\nabla\Psi) = u_m|\nabla\zeta|^2 \text{ in } \Omega, \quad \Psi = 0 \text{ on } \partial\Omega. \tag{4.13}$$

Since $a \in W^{1,\infty}(B(\sigma_0, R) \cap \Omega)$, and $\psi_m \in H^{3/2}(B(\sigma_0, R) \cap \partial\Omega))$, by the local regularity results of solutions to elliptic equations (see Section 2) we know that Ψ is in $H^2(B(\sigma_0, R)\cap\Omega)$ and therefore $\partial\Psi/\partial n$ is in $H^{1/2}(B(\sigma_0, R)\cap\partial\Omega)$. Moreover we have

$$\left\| a\frac{\partial\Psi}{\partial n} \right\|_{H^{1/2}(B(\sigma_0,R)\cap\partial\Omega)} \le c\|\Psi\|_{H^2(B(\sigma_0,R)\cap\Omega)} \le c\|u_m|\nabla\zeta|^2\|_{L^2(\Omega)}. \tag{4.14}$$

Multiplying equation (4.13) by u_m and integrating by parts we have:

$$\int_{\Omega} u_m(x)^2|\nabla\zeta(x)|^2\, dx = \int_{\Omega} a\nabla\Psi \cdot \nabla u_m\, dx - \int_{\partial\Omega} a\frac{\partial\Psi(\sigma)}{\partial n(\sigma)}\psi_m(\sigma)\, d\sigma$$

$$= -\int_{\partial\Omega} a\frac{\partial\Psi(\sigma)}{\partial n(\sigma)}\psi_m(\sigma)\, d\sigma,$$

where we have used the fact that since u_m satisfies (4.10) and $\Psi \in H_0^1(\Omega)$ we have $\int_{\Omega} a\nabla\Psi \cdot \nabla u_m\, dx = 0$. It follows that

$$\int_{\Omega} u_m(x)^2|\nabla\zeta(x)|^2\, dx \le c\left\| a\frac{\partial\Psi}{\partial n} \right\|_{H^{1/2}(B(\sigma_0,R)\cap\partial\Omega)} \|\psi_m\|_{H^{-1/2}(\partial\Omega)},$$

which, using (4.14) and the fact that $|\nabla\zeta|^2 \le c|\nabla\zeta|$, yields

$$\|u_m\nabla\zeta\|_{L^2}^2 \le c\|u_m|\nabla\zeta|^2\|_{L^2}\|\psi_m\|_{H^{-1/2}(\partial\Omega)} \le c\|u_m\nabla\zeta\|_{L^2}\|\psi_m\|_{H^{-1/2}(\partial\Omega)}.$$

Finally the estimate on ψ_m given by Lemma 4.1 together with inequality (4.12) prove that $\|\nabla u_m\|_{L^2(\omega)} \le c/m$ for $m \ge m_0$. Since $u_m = 0$ on $\partial\omega \cap \partial\Omega$, by Poincaré inequality we have $\|u_m\|_{L^2(\omega)} \le c(\omega, \Omega)\|\nabla u_m\|_{L^2(\omega)}$, and the lemma is proved. $\square$

Our next step will be to show that in the ball $B(\sigma_0, R)$ the solution u_m is not *too small*:

Lemma 4.3. *With the notations and assumptions of Lemma 4.2 there exist* $m_1 \geq m_0$ *and a constant* $c_0 > 0$ *depending on R such that for all* $m \geq m_1$

$$\int_{B(\sigma_0, R) \cap \Omega} |\nabla u_m(x)|^2 \, dx \geq c_0.$$

Proof. Recall that if we denote by $\gamma_0 : H^1(\Omega) \longrightarrow H^{1/2}(\partial\Omega)$ the trace operator $u \mapsto u_{|\partial\Omega}$, there exists a positive constant c such that for all $u \in H^1(\Omega)$ we have (see Section 2)

$$\|\gamma_0(u)\|^2_{H^{1/2}(\partial\Omega)} \leq c\big(\|\nabla u\|^2_2 + \|\gamma_0(u)\|^2_{L^2(\partial\Omega)}\big).$$

Applying this to u_m for $m \geq m_0$, one sees that since $\gamma_0(u_m) = \psi_m$ and $\|\psi_m\|_{L^2(\partial\Omega)} \leq cm^{-1/2}$ (cf. Lemma 4.1):

$$1 = \|\gamma_0(u_m)\|^2_{H^{1/2}(\partial\Omega)} \leq c\left(\|\nabla u_m\|^2_2 + \frac{c}{m}\right)$$

$$\leq c \int_{B(\sigma_0, R) \cap \Omega} |\nabla u_m(x)|^2 \, dx + \frac{c}{m^2} + \frac{c}{m}.$$

Clearly, upon taking $m_1 \geq m_0$ large enough and $m \geq m_1$, this yields the claimed result of the lemma. $\qquad\square$

We can now state and prove the first identification result we had in mind: namely that if $\Lambda_{a_0} = \Lambda_{a_1}$ then $a_0 = a_1$ on the boundary.

Theorem 4.4. *Let* Ω *be a bounded domain and* $\sigma_* \in \partial\Omega$ *such that* Ω *is* $C^{1,1}$ *in a neighborhood* $B(\sigma_*, R_*)$ *of* σ_*. *Assume that* $a_j \in L^\infty(\Omega) \cap W^{1,\infty}(\Omega \cap B(\sigma_*, R_*))$ *with* $a_j \geq \varepsilon_0 > 0$ *in* Ω. *Let* Λ_{a_j} *be as in (4.1) and assume that* $\Lambda_{a_0} = \Lambda_{a_1}$. *Then one has* $a_0 = a_1$ *on* $B(\sigma_*, R_*) \cap \partial\Omega$.

In particular if Ω *is* $C^{1,1}$ *and* $a_j \in W^{1,\infty}(\Omega)$, *the equality* $\Lambda_{a_0} = \Lambda_{a_1}$ *implies that* $a_0 = a_1$ *on* $\partial\Omega$.

Proof. If there exists $\sigma_0 \in B(\sigma_*, R_*) \cap \partial\Omega$ such that (for instance) $a_0(\sigma_0) > a_1(\sigma_0)$, then the a_j's being continuous in a neighborhood of σ_0, there exist $\beta > 0$ and $R > 0$ such that $B(\sigma_0, R) \subset B(\sigma_*, R_*)$ and

$$\forall x \in B(\sigma_0, R) \cap \Omega, \quad a_0(x) \geq a_1(x) + \beta.$$

Now denote by $u_{jm} \in H^1(\Omega)$ the solution of

$$\begin{cases} -\mathrm{div}\,(a_j \nabla u_{jm}) = 0 & \text{in } \Omega \\ u_{jm} = \psi_m & \text{on } \partial\Omega. \end{cases}$$

Denoting $B := B(\sigma_0, R) \cap \Omega$ we have

$$Q_{a_0}(\psi_m) = \int_\Omega a_0(x)|\nabla u_{0m}|^2\, dx \geq \int_B a_0(x)|\nabla u_{0m}|^2\, dx$$

$$\geq \int_B a_1(x)|\nabla u_{0m}|^2\, dx + \beta \int_B |\nabla u_{0m}|^2\, dx$$

$$\geq \int_B a_1(x)|\nabla u_{0m}|^2\, dx + \beta c_0$$

$$\geq \int_\Omega a_1(x)|\nabla u_{0m}|^2\, dx - \frac{c}{m^2} + \beta c_0,$$

provided $m \geq m_1$, where m_1 is given by Lemma 4.3 (in the last estimate we have used Lemma 4.2, that is $\int_{\Omega \setminus B} a_1(x)|\nabla u_{0m}(x)|^2\, dx \leq cm^{-2}$). It follows that since by definition of u_{1m} (we denote by K_{ψ_m} the convex set $K_{\psi_m} := \psi_m + H_0^1(\Omega) := \{v \in H^1(\Omega);\ v = \psi_m \text{ on } \partial\Omega\}$)

$$\int_\Omega a_1(x)|\nabla u_{0m}|^2\, dx \geq \min_{v \in K_{\psi_m}} \int_\Omega a_1(x)|\nabla v|^2\, dx$$

$$= \int_\Omega a_1(x)|\nabla u_{1m}|^2\, dx = Q_{a_1}(\psi_m),$$

if $m_2 \geq m_1$ is large enough so that for $m \geq m_2$ we have $cm^{-2} \leq \beta c_0/2$, we have

$$Q_{a_0}(\psi_m) \geq \int_\Omega a_1(x)|\nabla u_{0m}|^2\, dx + \frac{1}{2}\beta c_0 \geq Q_{a_1}(\psi_m) + \frac{1}{2}\beta c_0. \qquad (4.15)$$

Since $\Lambda_{a_0} = \Lambda_{a_1}$, we have $Q_{a_0}(\psi_m) = Q_{a_1}(\psi_m)$, which is in contradiction with the above inequality. Therefore we have $a_0 = a_1$ on $B(\sigma_*, R_*) \cap \partial\Omega$ and the proposition is proved. $\qquad\square$

In order to prove that the assumption $\Lambda_{a_0} = \Lambda_{a_1}$ implies that $a_0 = a_1$ at order one on the boundary, that is $a_0 = a_1$ and $\nabla a_0 = \nabla a_1$ on $\partial\Omega$, we need to show a little bit more than the estimate of Lemma 4.3. Indeed, since we know that $a_0 = a_1$ on the boundary, the tangential derivatives of a_0 and a_1 coincide and so it remains to show that $\partial a_0/\partial n = \partial a_1/\partial n$. If this were not so, we would find $\beta > 0$ such that in a neighborhood $B := B(\sigma_0, R) \cap \Omega$ of some $\sigma_0 \in \partial\Omega$:

$$\forall x \in B, \qquad a_0(x) \geq a_1(x) + \beta\,\mathrm{dist}(x, \partial\Omega).$$

(Here $\mathrm{dist}(x, \partial\Omega)$ denotes the distance of x to the boundary). After a detailed inspection of the above proof of Proposition 4.4, one is convinced that if we knew that $\int_B \mathrm{dist}(x, \partial\Omega)|\nabla u_{0m}|^2\, dx$ has a *greater magnitude* than $\int_{\Omega \setminus B} a_1(x)|\nabla u_{0m}|^2\, dx$ (which is $O(m^{-2})$), then the whole line of the above argument might be carried out, and we would obtain a contradiction to the fact that $\Lambda_{a_0} = \Lambda_{a_1}$, or equivalently $Q_{a_0} = Q_{a_1}$.

Remark 4.5. One can see through the above proof that the identification result is in some sense local. This means that if for all $\varphi \in H^{1/2}(\partial\Omega)$ with $\mathrm{supp}(\varphi) \subset B(\sigma, R_*) \cap \partial\Omega$ one has $\Lambda_{a_0}(\varphi) = \Lambda_{a_1}(\varphi)$, then $a_0 = a_1$ on $B(\sigma_*, R_*) \cap \partial\Omega$. $\qquad\square$

In order to obtain the estimate we are lacking for now, we shall use a special function equivalent to $\mathrm{dist}(x, \partial\Omega)$. Indeed let us denote by $\varphi_1 \in H_0^1(\Omega)$ the first eigenfunction of the operator $u \mapsto -\mathrm{div}(a\nabla u)$ with Dirichlet boundary condition, that is

$$-\mathrm{div}(a\nabla\varphi_1) = \lambda_1\varphi_1 > 0 \text{ in } \Omega, \quad \varphi_1 \in H_0^1(\Omega), \quad \int_\Omega \varphi_1(x)^2\,dx = 1. \quad (4.16)$$

It is a classical consequence of the Hopf maximum principle (see for instance D. Gilbarg and N. S. Trudinger [17]) that on a part $B(\sigma_0, R) \cap \partial\Omega$ of the boundary which is $C^{1,1}$ one has $\partial\varphi_1/\partial n \leq -c_*(R, \Omega) < 0$ and

$$\forall x \in B(\sigma_0, R) \cap \overline{\Omega}, \quad c_0\,\mathrm{dist}(x, \partial\Omega) \leq \varphi_1(x) \leq c_1\,\mathrm{dist}(x, \partial\Omega).$$

Lemma 4.6. *With the notations and assumptions of Lemma 4.2 there exist $m_1 \geq m_0$ and a constant $c_0 > 0$ depending on R such that for all $m \geq m_1$*

$$\int_{B(\sigma_0, R) \cap \Omega} \mathrm{dist}(x, \partial\Omega)|\nabla u_m(x)|^2\,dx \geq \frac{c_0}{m}.$$

Proof. Let φ_1 be as in (4.16). Then we know that for some positive constants c_*, c_{1*}, c_{2*} we have (here again $B := B(\sigma_0, R) \cap \Omega$)

$$\begin{cases} \forall \sigma \in B(\sigma_0, R) \cap \partial\Omega, & -\dfrac{\partial\varphi_1}{\partial n} \geq c_* > 0 \\ \forall x \in B, & c_{1*}\,\mathrm{dist}(x, \partial\Omega) \leq \varphi_1(x) \leq c_{2*}\,\mathrm{dist}(x, \partial\Omega). \end{cases} \quad (4.17)$$

Assuming that $m \geq m_0$ is large enough so that $\mathrm{supp}(\psi_m) \subset B(\sigma_0, R) \cap \partial\Omega$, and multiplying (4.10) by $u_m\varphi_1$, which belongs to $H_0^1(\Omega)$ since φ_1 is in $W^{1,\infty}(\Omega)$ and vanishes on $\partial\Omega$, we obtain

$$\int_\Omega a(x)|\nabla u_m|^2\varphi_1(x)\,dx + \int_\Omega a(x)u_m(x)\nabla u_m(x)\cdot\nabla\varphi_1(x)\,dx = 0. \quad (4.18)$$

But noting that $u_m\nabla u_m = \nabla(u_m^2)/2$, we may integrate by parts the second term of the above equality to see that (we use also the equation satisfied by φ_1):

$$\int_\Omega a(x)u_m(x)\nabla u_m(x)\cdot\nabla\varphi_1(x)\,dx = \frac{1}{2}\int_{\partial\Omega}\psi_m(\sigma)^2 a(\sigma)\frac{\partial\varphi_1(\sigma)}{\partial n(\sigma)}\,d\sigma$$
$$+ \frac{\lambda_1}{2}\int_\Omega u_m(x)^2\varphi_1(x)\,dx.$$

Inserting this into (4.18) yields:

$$\int_\Omega a(x)|\nabla u_m|^2\varphi_1(x)\,dx = -\frac{\lambda_1}{2}\int_\Omega u_m(x)^2\varphi_1(x)\,dx$$
$$-\frac{1}{2}\int_{\partial\Omega}\psi_m(\sigma)^2 a(\sigma)\frac{\partial\varphi_1}{\partial n(\sigma)}\,d\sigma. \qquad (4.19)$$

Now we may use the Hopf maximum principle expressed in the first inequality of (4.17) to see that for some positive constant c, due to property (iii) of Lemma 4.1 with $s = 0$, we have

$$-\int_{\partial\Omega}\psi_m(\sigma)^2 a(\sigma)\frac{\partial\varphi_1}{\partial n(\sigma)}\,d\sigma \geq c_*\int_{\partial\Omega}a(\sigma)\psi_m(\sigma)^2\,d\sigma \geq \frac{c}{m}. \qquad (4.20)$$

As far as the first term in the right hand side of (4.19) is concerned, observe that if $\Psi \in H_0^1(\Omega)$ solves

$$-\mathrm{div}(a\nabla\Psi) = u_m\varphi_1 \text{ in } \Omega, \quad \Psi = 0 \text{ on } \partial\Omega,$$

after multiplying this equation by u_m and integrating by parts, we obtain (because u_m satisfies (4.10) and so we have $\int_\Omega a\nabla u_m\cdot\nabla\Psi\,dx = 0$)

$$\int_\Omega u_m(x)^2\varphi_1(x)\,dx = -\int_{\partial\Omega}\psi_m(\sigma)a(\sigma)\frac{\partial\Psi(\sigma)}{\partial n(\sigma)}\,d\sigma$$

$$\leq c\|\psi_m\|_{H^{-1/2}(\partial\Omega)}\left\|a\frac{\partial\Psi}{\partial n}\right\|_{H^{1/2}(B(\sigma_0,R)\cap\partial\Omega)} \qquad (4.21)$$

$$\leq c\|\psi_m\|_{H^{-1/2}(\partial\Omega)}\|u_m\varphi_1^{1/2}\|_2,$$

where we have used the fact that $\Psi \in H^2(B)$ with $B := B(\sigma_0, R) \cap \Omega$, and

$$\left\|a\frac{\partial\Psi}{\partial n}\right\|_{H^{1/2}(B(\sigma_0,R)\cap\partial\Omega)} \leq c\|\Psi\|_{H^2(B)} \leq c\|u_m\varphi_1\|_2 \leq c\|u_m\varphi_1^{1/2}\|_2,$$

thanks to classical regularity theorems (see Proposition 2.3) and the fact that $\varphi_1 \leq c\varphi_1^{1/2}$. Finally from (4.21) we obtain that $\|u_m\varphi_1^{1/2}\|_2 \leq c\|\psi_m\|_{H^{-1/2}(\partial\Omega)}$ and using Lemma 4.1 with $s = -1/2$, the latter estimate shows that

$$\int_\Omega u_m(x)^2\varphi_1(x)\,dx \leq c\|\psi_m\|_{H^{-1/2}(\partial\Omega)}^2 \leq \frac{c}{m^2}. \qquad (4.22)$$

It is now plain that plugging estimates (4.20) and (4.22) into (4.19) finishes the proof of the lemma. □

At this point we can prove the following identification result.

Theorem 4.7. *Let Ω be a bounded domain and $\sigma_* \in \partial\Omega$ such that Ω is $C^{1,1}$ in a neighborhood $B(\sigma_*, R_*)$ of σ_*. Assume that $a_j \in L^\infty(\Omega) \cap C^1(\overline{\Omega} \cap B(\sigma_*, R_*))$ with $a_j \geq \varepsilon_0 > 0$ in Ω. Let Λ_{a_j} be as in (4.1) and assume that $\Lambda_{a_0} = \Lambda_{a_1}$. Then one has*

$a_0 = a_1$ and $\nabla a_0 = \nabla a_1$ on $B(\sigma_*, R_*) \cap \partial\Omega$.

In particular if Ω is $C^{1,1}$ and $a_j \in C^1(\overline{\Omega})$, the equality $\Lambda_{a_0} = \Lambda_{a_1}$ implies that $a_0 = a_1$ and $\nabla a_0 = \nabla a_1$ on $\partial\Omega$.

Proof. By Proposition 4.7 we know that $a_0 = a_1$ on $B(\sigma_*, R_*) \cap \partial\Omega$, therefore in order to prove that $\nabla a_0 = \nabla a_1$ on this portion of the boundary, it is enough to show that $\partial a_0/\partial n = \partial a_1/\partial n$. If there exists $\sigma_0 \in B(\sigma_*, R_*) \cap \partial\Omega$ such that $\partial a_0(\sigma_0)/\partial n(\sigma_0) \neq \partial a_1(\sigma_0)/\partial n(\sigma_0)$, then the a_j's being C^1 and equal on $B(\sigma_*, R_*) \cap \partial\Omega$, there exist $\beta > 0$ and $R > 0$ such that $B(\sigma_0, R) \subset B(\sigma_*, R_*)$ and (for instance)

$$\forall x \in B(\sigma_0, R) \cap \Omega, \quad a_0(x) \geq a_1(x) + \beta \operatorname{dist}(x, \partial\Omega).$$

We denote again by u_{jm} the solution of $\operatorname{div}(a_j \nabla u_{jm}) = 0$ and $u_{jm} = \psi_m$ on $\partial\Omega$, and by $\varphi_1 \in H_0^1(\Omega)$ the eigenfunction considered in (4.16) with $a := a_0$. If $B := B(\sigma_0, R) \cap \Omega$ we have (here we use Lemma 4.6)

$$\int_\Omega a_0(x)|\nabla u_{0m}|^2 \, dx \geq \int_B a_0(x)|\nabla u_{0m}|^2 \, dx$$

$$\geq \int_B a_1(x)|\nabla u_{0m}|^2 \, dx + \beta \int_B \operatorname{dist}(x, \partial\Omega)|\nabla u_{0m}|^2 \, dx$$

$$\geq \int_B a_1(x)|\nabla u_{0m}|^2 \, dx + \frac{\beta c_0}{m}$$

$$\geq \int_\Omega a_1(x)|\nabla u_{0m}|^2 \, dx - \frac{c}{m^2} + \frac{\beta c_0}{m},$$

provided $m \geq m_1$, given by Lemma 4.6 (in the last estimate we have used Lemma 4.2). It follows that since by definition of u_{1m}

$$\int_\Omega a_1(x)|\nabla u_{0m}|^2 \, dx \geq Q_{a_1}(\psi_m),$$

for $m_2 \geq m_1$ large enough so that for $m \geq m_2$ we have $cm^{-1} \leq \beta c_0/2$,

$$Q_{a_0}(\psi_m) \geq \int_\Omega a_1(x)|\nabla u_{0m}|^2 \, dx + \frac{\beta c_0}{2m} \geq Q_{a_1}(\psi_m) + \frac{\beta c_0}{2m}.$$

This is a contradiction with the fact that $Q_{a_0}(\psi_m) = Q_{a_1}(\psi_m)$. Therefore we have $\partial a_0/\partial n = \partial a_1/\partial n$ on $B(\sigma_*, R_*) \cap \partial\Omega$ and the theorem is proved. □

Remark 4.8. It is interesting to note that the arguments used to prove Theorem 4.4 may be used to give a stability estimate of the type

$$\|a_0 - a_1\|_{L^\infty(\partial\Omega)} \leq c(M)\|\Lambda_{a_0} - \Lambda_{a_1}\|_{\mathcal{L}(H^{1/2}(\partial\Omega), H^{-1/2}(\partial\Omega))}, \tag{4.23}$$

for all $a_j \in W^{1,\infty}(\Omega)$ such that $\|a_j\|_{W^{1,\infty}(\Omega)} \leq M$, where $c(M)$ is a constant depending only on Ω and M. Indeed assume that $a_0, a_1 \in W^{1,\infty}(\Omega)$ are such that $a_0 - a_1 \not\equiv 0$ on the boundary and that if $2\beta := \|a_0 - a_1\|_{L^\infty(\partial\Omega)}$, for some $\sigma_0 \in \partial\Omega$ and $R > 0$

we have

$$\forall x \in B(\sigma_0, R) \cap \Omega, \quad a_0(x) \geq a_1(x) + \beta.$$

Then according to (4.15) for some constant $c_0 > 0$ we have that

$$\frac{1}{2}\beta c_0 \leq Q_{a_0}(\psi_m) - Q_{a_1}(\psi_m) = \langle (\Lambda_{a_0} - \Lambda_{a_1})\psi_m, \psi_m \rangle$$

$$\leq \|\Lambda_{a_0} - \Lambda_{a_1}\|_{\mathcal{L}(H^{1/2}(\partial\Omega), H^{-1/2}(\partial\Omega))}$$

since $\|\psi_m\|_{H^{1/2}(\partial\Omega)} = 1$. It follows that (4.23) holds with $c(M) := 4/c_0$.

5 Identification of potentials in Schrödinger operators

Our third lecture will treat a few more general identification problems for elliptic operators, through appropriate boundary measurements. For $a \in \mathcal{A}_{ad}$ and $q \in L^\infty(\Omega)$ denote by $L := L_{a,q}$ the (formal) elliptic operator

$$Lu := L_{a,q}u := -\mathrm{div}(a\nabla u) + qu. \tag{5.1}$$

It is known that (see Section 2) the subspace $N(L)$ or $N(L_{a,q})$ defined to be

$$N(L_{a,q}) := \{\varphi \in H_0^1(\Omega); \ L_{a,q}\varphi = 0 \text{ in } H^{-1}(\Omega)\}, \tag{5.2}$$

is finite dimensional. Then we introduce the following way of saying:

Definition 5.1. We shall say that zero is not an eigenvalue of $L_{a,q}$ if $N(L_{a,q}) = \{0\}$.

If zero is not an eigenvalue of $L_{a,q}$ then for any given $\varphi \in H^{1/2}(\partial\Omega)$ there exists a unique $u \in H^1(\Omega)$ such that

$$\begin{cases} -\mathrm{div}(a\nabla u) + qu = 0 & \text{in } \Omega \\ u = \varphi & \text{on } \partial\Omega, \end{cases} \tag{5.3}$$

and we may define the corresponding Poincaré–Steklov operator $\Lambda_{a,q}$

$$\Lambda_{a,q}(\varphi) = a\frac{\partial u}{\partial n}, \qquad \Lambda_{a,q} : H^{1/2}(\partial\Omega) \longrightarrow H^{-1/2}(\partial\Omega). \tag{5.4}$$

Now the question we may ask is the following: if for two sets of coefficients a_j, q_j (where $j = 0, 1$) we have $\Lambda_{a_0,q_0} = \Lambda_{a_1,q_1}$ can we conclude that $a_0 = a_1$ and $q_0 = q_1$?

Before trying to answer this question let us show that $\Lambda_{a,q}$ is a continuous selfadjoint operator on $H^{1/2}(\partial\Omega)$.

Lemma 5.2. *For given $a \in \mathcal{A}_{ad}$ and $q \in L^\infty(\Omega)$ assume that zero is not an eigenvalue of $\Lambda_{a,q}$ defined by (5.4). Then $\Lambda_{a,q}$ is a selfadjoint operator mapping $H^{1/2}(\partial\Omega)$ into $H^{-1/2}(\partial\Omega)$.*

Proof. The fact that $\Lambda_{a,q}$ is continuous from $H^{1/2}(\partial\Omega)$ into $H^{-1/2}(\partial\Omega)$ is a consequence of Lemma 2.2. In order to see that $\Lambda_{a,q}$ is selfadjoint, let u solve (5.3) and for $\psi \in H^{1/2}(\partial\Omega)$ let $v \in H^1(\Omega)$ solve

$$-\mathrm{div}\,(a\nabla v) + qv = 0 \text{ in } \Omega, \quad v = \psi \text{ on } \partial\Omega.$$

Upon multiplying this equation by u, equation (5.3) by v and integrating by parts, one sees that:

$$\int_\Omega a(x)\nabla v(x) \cdot \nabla u(x)\,dx + \int_\Omega q(x)v(x)u(x)\,dx = \int_{\partial\Omega} a(\sigma)\frac{\partial v(\sigma)}{\partial n(\sigma)}\varphi(\sigma)\,d\sigma$$

$$\int_\Omega a(x)\nabla u(x) \cdot \nabla v(x)\,dx + \int_\Omega q(x)u(x)v(x)\,dx = \int_{\partial\Omega} a(\sigma)\frac{\partial u(\sigma)}{\partial n(\sigma)}\psi(\sigma)\,d\sigma,$$

which yields that $\langle \Lambda_{a,q}\psi, \varphi\rangle = \langle \Lambda_{a,q}\varphi, \psi\rangle$, and hence $\Lambda_{a,q}^* = \Lambda_{a,q}$. □

This gives the following important identity:

Corollary 5.3. *If $a_j \in \mathcal{A}_{\mathrm{ad}}$ and $q_j \in L^\infty(\Omega)$ (for $j = 0, 1$) are such that zero is not an eigenvalue of Λ_{a_j,q_j}, then for all $\varphi_j \in H^{1/2}(\partial\Omega)$ if $u_j \in H^1(\Omega)$ solves (5.3) with $a := a_j$ and $q := q_j$ and $\varphi := \varphi_j$, we have:*

$$\langle (\Lambda_{a_0,q_0} - \Lambda_{a_1,q_1})\varphi_0, \varphi_1\rangle = \int_\Omega (a_0(x) - a_1(x))\nabla u_0(x) \cdot \nabla u_1(x)\,dx$$

$$+ \int_\Omega (q_0(x) - q_1(x))u_0(x)u_1(x)\,dx. \qquad (5.5)$$

Proof. Multiplying by u_1 the equation satisfied by u_0, and vice versa, integrating by parts gives:

$$\langle \Lambda_{a_0,q_0}\varphi_0, \varphi_1\rangle = \int_\Omega a_0(x)\nabla u_0(x) \cdot \nabla u_1(x)\,dx + \int_\Omega q_0(x)u_0(x)u_1(x)\,dx$$

$$\langle \Lambda_{a_1,q_1}\varphi_1, \varphi_0\rangle = \int_\Omega a_1(x)\nabla u_1(x) \cdot \nabla u_0(x)\,dx + \int_\Omega q_1(x)u_1(x)u_0(x)\,dx.$$

Subtracting the second equality from the first, and using the fact that according to Lemma 5.2 one has

$$\langle \Lambda_{a_1,q_1}\varphi_1, \varphi_0\rangle = \langle \Lambda_{a_1,q_1}\varphi_0, \varphi_1\rangle,$$

we see that identity (5.5) is proved. □

This corollary, although quite simple to prove, is of the utmost importance in the study of inverse or identification problems. Indeed it shows that identification of the coefficients a_j, q_j boils down to the proof of an abstract density result. Indeed denote by $E_0 := E_0(a_0, q_0, a_1, q_1)$ the space

$$E_0 := \operatorname{span}\{(\nabla u_0 \cdot \nabla u_1, u_0 u_1); \ u_j \in H^1(\Omega), \ L_{a_j, q_j} u_j = 0\}, \qquad (5.6)$$

in $L^1(\Omega) \times L^1(\Omega)$. Then if one could prove that E_0 is dense in $L^1(\Omega) \times L^1(\Omega)$, one could conclude that if $\Lambda_{a_0, q_0} = \Lambda_{a_1, q_1}$ then necessarily $a_0 = a_1$ and $q_0 = q_1$: indeed by identity (5.5) we would infer that $(a_0 - a_1, q_0 - q_1)$ is orthogonal to E_0, and the density of this space in $L^1(\Omega) \times L^1(\Omega)$ would imply that $(a_0 - a_1, q_0 - q_1) = (0, 0)$ in $L^\infty(\Omega) \times L^\infty(\Omega)$. More precisely, rather than the density of E_0 in $L^1(\Omega) \times L^1(\Omega)$, since E_0 depends on a_0, q_0, a_1, q_1, we need to know that if

$$(a_0 - a_1, q_0 - q_1) \in E_0(a_0, a_1, q_0, q_1)^\perp,$$

then $(a_0, q_0) = (a_1, q_1)$. At this point we are not able to show such a general result for any two pairs of coefficients (a_j, q_j). However we may state the following:

Theorem 5.4. *For $a_j \in \mathcal{A}_{\mathrm{ad}}$ and $q_j \in L^\infty(\Omega)$ so that zero is not an eigenvalue of L_{a_j, q_j}, define the spaces*

$$V_0(q_0, q_1) := \operatorname{span}\{u_0 u_1; \ u_j \in H^1(\Omega), \ L_{a_j, q_j} u_j = 0\}$$

$$V_1(a_0, a_1) := \operatorname{span}\{\nabla u_0 \cdot \nabla u_1; \ u_j \in H^1(\Omega), \ L_{a_j, q_j} u_j = 0\}.$$

Then if $a_0 = a_1$ and if $V_0(q_0, q_1)$ is dense in $L^1(\Omega)$, the equality $\Lambda_{a_0, q_0} = \Lambda_{a_0, q_1}$ implies that $q_0 = q_1$ in Ω.

Analogously if $q_0 = q_1$ and if $V_1(a_0, a_1)$ is dense in $L^1(\Omega)$, the equality $\Lambda_{a_0, q_0} = \Lambda_{a_1, q_0}$ implies that $a_0 = a_1$ in Ω.

Proof. Indeed if for instance $a_0 = a_1$ and $\Lambda_{a_0, q_0} = \Lambda_{a_0, q_1}$ then thanks to identity (5.5) for any $\varphi_j \in H^{1/2}(\partial\Omega)$ we have

$$\int_\Omega (q_0(x) - q_1(x)) u_0(x) u_1(x) \, dx = 0.$$

This means that $q_0 - q_1$ is orthogonal to $V_0(q_0, q_1)$, and since the latter is dense in $L^1(\Omega)$ we conclude that $q_0 = q_1$. The other claim of the theorem is seen to be true in an analogous manner. $\qquad\square$

We observe now that a practical way to prove that some coefficients may be identified in inverse problems, would be to prove that the spaces V_0 or V_1 are dense in $L^1(\Omega)$. In these lectures, due to lack of time we shall not consider this general problem (see [22] for a detailed treatment), but rather consider the simpler case $a_0 = a_1 \equiv 1$. The following results are essentially due to J. Sylvester and G. Uhlmann [38], although in part we shall use a simpler proof of one of their arguments, thanks to a simplification given by P. Hähner [18].

We begin with a result which says that there is a direct connection between the study of the general case and the special case $a_0 = a_1 \equiv 1$ and $q_0, q_1 \in L^\infty(\Omega)$.

Proposition 5.5. *Let* $a_j \in \mathcal{A}_{ad} \cap W^{2,\infty}(\Omega)$ *and* $q_j \in L^\infty(\Omega)$ *(for* $j = 0, 1$*) be such that zero is not an eigenvalue of* L_{a_j,q_j}*. Define*

$$\tilde{q}_j := \frac{q_j}{a_j} + \frac{\Delta a_j^{1/2}}{a_j^{1/2}}$$

and assume that $a_0 = a_1$ *and* $\nabla a_0 = \nabla a_1$ *on the boundary* $\partial\Omega$*. Then*

$$\Lambda_{a_0,q_0} = \Lambda_{a_1,q_1} \iff \Lambda_{1,\tilde{q}_0} = \Lambda_{1,\tilde{q}_1}.$$

Proof. For a given $\varphi \in H^{1/2}(\partial\Omega)$, if $u_j \in H^1(\Omega)$ satisfies

$$-\text{div}(a_j \nabla u_j) + q_j u_j = 0 \text{ in } \Omega, \quad u_j = \varphi \text{ on } \partial\Omega,$$

then using the Liouville's transform, that is setting $v_j := a_j^{1/2} u_j$, or equivalently $u_j = a_j^{-1/2} v$, one checks that

$$\text{div}(a_j \nabla u_j) = \text{div}\left(a_j^{1/2} \nabla v_j - \frac{\nabla a_j}{2a_j^{1/2}} v_j\right) = \text{div}\left(a_j^{1/2} \nabla v_j - v_j \nabla(a_j^{1/2})\right)$$

$$= a_j^{1/2}\left(\Delta v_j - \frac{\Delta a_j^{1/2}}{a_j^{1/2}} v_j\right).$$

Hence v_j satisfies

$$-\Delta v_j + \left(\frac{q_j}{a_j} + \frac{\Delta a_j^{1/2}}{a_j^{1/2}}\right) v_j = 0 \text{ in } \Omega, \quad v_j = a_j^{1/2}\varphi \text{ on } \partial\Omega.$$

Observe also that since v_j satisfies the above equation, it follows that

$$\Lambda_{1,\tilde{q}_j}(a_j^{1/2}\varphi) = \frac{\partial v_j}{\partial n} = a_j^{1/2}\frac{\partial u_j}{\partial n} + \frac{1}{2}a_j^{-1/2}\frac{\partial a_j}{\partial n}\varphi$$

$$= a_j^{-1/2}\Lambda_{a_j,q_j}(\varphi) + \frac{1}{2}a_j^{-1/2}\frac{\partial a_j}{\partial n}\varphi \tag{5.7}$$

Since $a_0 = a_1$ and $\nabla a_0 = \nabla a_1$ on $\partial\Omega$, and due to the fact that $\varphi \mapsto a_j^{1/2}\varphi$ is a homeomorphism on $H^{1/2}(\partial\Omega)$, we conclude that saying $\Lambda_{a_0,q_0} = \Lambda_{a_1,q_1}$ is equivalent to say that $\Lambda_{1,\tilde{q}_0} = \Lambda_{1,\tilde{q}_1}$. $\qquad\square$

Corollary 5.6. *Let* Ω *be a bounded* $C^{1,1}$ *domain. Assume that for* $q_j \in L^\infty(\Omega)$ *(with* $j = 0, 1$*) such that zero is not an eigenvalue of* L_{1,q_j} *we have the following property:*

$$\Lambda_{1,q_0} = \Lambda_{1,q_1} \implies q_0 = q_1 \text{ in } \Omega. \tag{5.8}$$

Then if $a_j \in \mathcal{A}_{ad} \cap W^{2,\infty}(\Omega)$ *are such that* $\Lambda_{a_0,0} = \Lambda_{a_1,0}$*, we have* $a_0 = a_1$ *in* $\overline{\Omega}$*.*

Proof. First according to Theorem 4.4 and Theorem 4.7, we know that whenever $\Lambda_{a_0,0} = \Lambda_{a_1,0}$, then $a_0 = a_1$ and $\nabla a_0 = \nabla a_1$ on the boundary $\partial\Omega$. Next setting

$$\tilde{q}_j := \frac{\Delta a_j^{1/2}}{a_j^{1/2}},$$

then clearly zero is not an eigenvalue of $L_{1,\tilde{q}_j}$, because saying that $v \in H_0^1(\Omega)$ satisfies $-\Delta v + \tilde{q}_j v = 0$ is equivalent to say that $\text{div}(a_j \nabla u) = 0$ with $u := a^{-1/2} v \in H_0^1(\Omega)$, and the latter equality implies $u = 0 = v$. Now Proposition 5.5 shows that if $\Lambda_{a_0,0} = \Lambda_{a_1,0}$, since a_0 and a_1, as well as ∇a_0 and ∇a_1, are equal on the boundary, we have $\Lambda_{1,\tilde{q}_0} = \Lambda_{1,\tilde{q}_1}$. Property (5.8) implies that $\tilde{q}_0 = \tilde{q}_1$ in Ω, but it is clear that if we set $q := \tilde{q}_j$, the functions $a_j^{1/2}$ (which are also in $H^1(\Omega)$) satisfy the elliptic equation

$$\begin{cases} -\Delta a_j^{1/2} + q a_j^{1/2} = 0 & \text{in } \Omega \\ a_j = a_0 & \text{on } \partial\Omega. \end{cases}$$

Since zero is not an eigenvalue of $L_{1,q}$ it follows that the above equation has a unique solution, that is $a_0 = a_1$ in Ω, and the corollary is proved. $\square$

Now our objective will be to prove Property (5.8), which is true when the dimension $N \geq 3$, and its proof is due to J. Sylvester and G. Uhlmann [38]. We state the following density theorem which will be proved in the next section.

Theorem 5.7. *Assume that $\Omega \subset \mathbb{R}^N$ is a bounded Lipschitz domain with $N \geq 3$. If $q_j \in L^\infty(\Omega)$ are such that zero is not an eigenvalue of L_{1,q_j}, then the space*

$$V_0(q_0, q_1) := \text{span}\{u_0 u_1; \ u_j \in H^1(\Omega) \text{ and } -\Delta u_j + q_j u_j = 0 \text{ in } \Omega\} \quad (5.9)$$

is dense in $L^1(\Omega)$. As a consequence if the Poincaré–Steklov operators Λ_{1,q_0} and Λ_{1,q_1} are equal, then $q_0 = q_1$ in Ω.

Remark 5.8. The case of dimension $N = 2$ is particular in two respects. First by a result of A.I. Nachman [33], which uses a completely different approach, one knows that if a_0 and a_1 are smooth enough then the equality of the Poincaré–Steklov operators $\Lambda_{a_0,0} = \Lambda_{a_1,0}$ implies that $a_0 = a_1$ in $\overline{\Omega}$. Using this result one may show that if q_0, q_1 are such that the first eigenvalue of the operator $(L_{1,q_j}, D(L_{1,q_j}))$ is positive, then the fact that $\Lambda_{1,q_0} = \Lambda_{1,q_1}$ implies $q_0 = q_1$.

However the case of dimension two is not thoroughly well understood: it is not known whether for a general pair $q_0, q_1 \in L^\infty(\Omega)$ the space $V_0(q_0, q_1)$ is dense in $L^1(\Omega)$, nor is it known whether the functions q_0, q_1 must be equal whenever $\Lambda_{1,q_0} = \Lambda_{1,q_1}$. Due to lack of time unfortunately we will not treat the case $N = 2$ in these lectures.

6 Density of products of solutions

In our fourth and last lecture we prove the density result invoked at the end of the previous lecture. In order to establish the density Theorem 5.7 the main idea, as pointed out in the original paper of J. Sylvester and G. Uhlmann [38], is the following. Since the exponential functions $\exp((i\xi \pm \eta) \cdot x)$ used by A.P. Calderón (see Section 3, in particular (3.7) and (3.8)) are such that the space spanned by their product is dense, each of them being a harmonic function, one may consider the operators $u \mapsto -\Delta u + q_j u$ as perturbations of the Laplacian and prove that the space $E(q_0, q_1)$ defined in (5.9) is dense. The point is that this clever idea works, at least when $N \geq 3$. However here we will follow the point of view presented in a later work by P. Hähner [18].

In what follows we consider a potential $q \in L^\infty(\Omega)$ such that zero is not an eigenvalue of the operator $L_{1,q}$. We wish to construct solutions of

$$u \in H^1(\Omega), \quad -\Delta u + qu = 0 \tag{6.1}$$

such that

$$u(x) = e^{\zeta \cdot x}(1 + \psi(x)), \quad \text{with } \zeta = i\xi + \eta,$$

and

$$\xi, \eta \in \mathbb{R}^N, \quad |\xi| = |\eta|, \quad \xi \cdot \eta = 0. \tag{6.2}$$

Since $L_{1,q}u = e^{\zeta \cdot x}[-\Delta \psi - 2\zeta \cdot \nabla \psi + q(1 + \psi)]$, in order for u to be a solution of (6.1), the new unknown function ψ should be a solution of

$$\psi \in H^1(\Omega), \quad -\Delta \psi - 2\zeta \cdot \nabla \psi + q\psi = -q. \tag{6.3}$$

Moreover it is desirable to have ψ *small* in some appropriate norm, at least when $|\zeta| \to +\infty$, so that u is close to $e^{(i\xi+\eta) \cdot x}$. We are going to solve (6.3) when $\Omega \subset (-R, R)^N$ for some $R > 0$ and in the special case $\zeta = se_1 + i\xi$ for some $s \in \mathbb{R}$, $s \neq 0$, and $\xi \in \mathbb{R}^N$ such that $e_1 \cdot \xi = 0$. Note that this is not any restriction since the operators are essentially invariant under translations and rotations of $\mathbb{R}^N$; also we should say that we denote $e_1 := (1, 0, \ldots, 0)$, and analogously e_k will be the k-th vector of the standard basis of $\mathbb{R}^N$. We denote by Q the cube $(-R, R)^N$ and we introduce

$$Z_0 := \left\{ \alpha \in \mathbb{R}^N; \; \frac{\alpha_1 R}{\pi} - \frac{1}{2} \in \mathbb{Z}, \; \frac{\alpha_j R}{\pi} \in \mathbb{Z} \text{ for } j \geq 2 \right\}. \tag{6.4}$$

Observe that when $\alpha \in Z_0$ then $\alpha_1 \neq 0$, more precisely $|\alpha_1| \geq \pi/2R$ and, as we shall see below, this is the reason for the choice of a shifted grid. Also note that we may write

$$Z_0 = \frac{\pi}{2R}e_1 + \frac{\pi}{R}\mathbb{Z}^N. \tag{6.5}$$

We recall here the classical definition of Q-periodic functions:

Definition 6.1. A function $u \in H^1_{\text{loc}}(\mathbb{R}^N)$ is said to be Q-periodic if for all $1 \leq k \leq N$ it satisfies $u(\cdot + 2Re_k) = u(\cdot)$. Such a function is completely determined by its values on Q and the subspace of Q-periodic functions of $H^1_{\text{loc}}(\mathbb{R}^N)$ is denoted by $H^1_{\text{per}}(Q)$.

Equivalently one can give a local definition of $H^1_{\text{per}}(Q)$. A function $u \in H^1(Q)$ is said to belong to $H^1_{\text{per}}(Q)$, or to be Q-periodic, if for all $1 \leq k \leq N$ we have $u(\sigma + 2Re_k) = u(\sigma)$ for $\sigma \in \partial Q$ (since the trace $u_{|\partial Q}$ is defined, the condition makes sense). In this case, a Q-periodic function in the latter sense can be extended by periodicity to all $\mathbb{R}^N$, and one may check that the two definitions of $H^1_{\text{per}}(Q)$ coincide.

The space $H^1_{\text{per}}(Q)$ is a closed subspace of $H^1(Q)$ and it is endowed with the natural topology of $H^1(Q)$. Analogously the space $H^2_{\text{per}}(Q)$ is defined to be:

$$H^2_{\text{per}}(Q) := \left\{ u \in H^1_{\text{per}}(Q); \ \partial_k u \in H^1_{\text{per}}(Q) \text{ for } 1 \leq k \leq N \right\}.$$

Next we define the notion of Z_0-quasiperiodic functions.

Definition 6.2. A function $u \in H^1_{\text{loc}}(\mathbb{R}^N)$ is said to be Z_0-quasiperiodic if the function $x \mapsto \exp(-i\pi x_1/2R)u(x)$ is Q-periodic. The space of Z_0-periodic functions which are in $H^1_{\text{loc}}(\mathbb{R}^N)$ will be denoted by $H^1_{Z_0}(Q)$. This is a closed subspace of $H^1(Q)$.

One may check that this is a closed subspace of $H^1(Q)$. Analogously the space $H^2_{Z_0}(Q)$ is the space of Z_0-quasiperiodic functions which are in $H^2_{\text{loc}}(\mathbb{R}^N)$. Equivalently, the space $H^1_{Z_0}(Q)$ (resp. $H^2_{Z_0}(Q)$) can be defined as the space generated by $\exp(i\pi x_1/2R)\varphi$ for $\varphi \in H^1_{\text{per}}(Q)$ (resp. $\varphi \in H^2_{\text{per}}(Q)$).

For $\alpha \in Z_0$ we denote by φ_α the function

$$\varphi_\alpha(x) := (2R)^{-N/2} e^{i\alpha \cdot x}. \tag{6.6}$$

One checks easily that each φ_α is Z_0-quasiperiodic, that $(\varphi_\alpha)_{\alpha \in Z_0}$ is a Hilbert basis of $L^2(Q)$ and that for any $\alpha \in \mathbb{Z}^N$

$$\nabla \varphi_\alpha = i\alpha \varphi_\alpha, \qquad \Delta \varphi_\alpha = -|\alpha|^2 \varphi_\alpha.$$

A useful observation about Q-periodic functions is that when $u, v \in C^1(\overline{Q}) \cap H^1_{\text{per}}(Q)$ then we have $\int_{\partial Q} u(\sigma)\overline{v(\sigma)}n_k(\sigma)\,d\sigma = 0$ (because $\sigma \mapsto n_k(\sigma)$ is anti-periodic on ∂Q) and therefore $\int_Q \nabla u(x)\,\overline{v(x)}\,dx = -\int_Q u(x)\,\nabla \overline{v(x)}\,dx$. Also for $u, v \in C^2(\overline{Q}) \cap H^2_{\text{per}}(Q)$, since $\sigma \mapsto \nabla u(\sigma) \cdot n(\sigma)$ is anti-periodic, we have

$$\int_{\partial Q} \frac{\partial u(\sigma)}{\partial n(\sigma)} \overline{v(\sigma)}\,d\sigma = 0, \quad \text{and} \quad \int_{\partial Q} \left(\frac{\partial u(\sigma)}{\partial n(\sigma)} \overline{v(\sigma)} - u(\sigma) \frac{\overline{\partial v(\sigma)}}{\partial n(\sigma)} \right) d\sigma = 0,$$

and hence

$$\int_Q \Delta u(x)\,\overline{v(x)}\,dx = \int_Q u(x)\Delta \overline{v(x)}\,dx.$$

For our study of equation (6.3) it is useful to point out that if $u, v \in C^2(\overline{Q})$ are Z_0-quasiperiodic, then setting

$$\varphi(x) := \exp(-i\pi x_1/2R)u(x), \quad \psi(x) := \exp(-i\pi x_1/2R)v(x),$$

we have two Q-periodic functions and

$$\nabla u = \exp(i\pi x_1/2R)\nabla \varphi(x) + \frac{i\pi}{2R}\exp(i\pi x_1/2R)\varphi(x)e_1$$

$$\nabla v = \exp(i\pi x_1/2R)\nabla \psi(x) + \frac{i\pi}{2R}\exp(i\pi x_1/2R)\psi(x)e_1.$$

Hence one can easily verify that

$$\int_{\partial Q}\left(\frac{\partial u(\sigma)}{\partial n(\sigma)}\overline{v(\sigma)} - u(\sigma)\frac{\overline{\partial v(\sigma)}}{\partial n(\sigma)}\right)d\sigma = \int_{\partial Q}\left(\frac{\partial \varphi(\sigma)}{\partial n(\sigma)}\overline{\psi(\sigma)} - \varphi(\sigma)\frac{\overline{\partial \psi(\sigma)}}{\partial n(\sigma)}\right)d\sigma$$

$$+ \frac{i\pi}{R}\int_{\partial\Omega}\varphi(\sigma)\overline{\psi(\sigma)}\,e_1 \cdot n(\sigma)\,d\sigma$$

$$= 0.$$

Therefore for any $u, v \in C^2(\overline{Q})$ which are Z_0-quasiperiodic we have

$$\begin{cases} \int_Q \nabla u(x)\,\overline{v(x)}\,dx = -\int_Q u(x)\,\overline{\nabla v(x)}\,dx \\ \int_Q \Delta u(x)\,\overline{v(x)}\,dx = \int_Q u(x)\overline{\Delta v(x)}\,dx. \end{cases} \tag{6.7}$$

Invoking a density argument, one is convinced that the above relations are still valid when the derivatives exist in the weak sense (the first equality in (6.7) holds for $u, v \in H^1_{Z_0}(Q)$, while the second one is valid for $u, v \in H^2_{Z_0}(Q)$ which may be defined in much the same way as $H^1_{Z_0}(Q)$). We are now in a position to solve an elliptic equation which corresponds to the differential part of equation (6.3).

Proposition 6.3. *Let $s \in \mathbb{R}$, $s \neq 0$ and $\xi \in \mathbb{R}^N$ such that $\xi \cdot e_1 = 0$ and set $\zeta := i\xi + se_1$. Then for any $f \in L^2(Q)$ there exists a unique $\psi \in H^1_{Z_0}(Q) \cap H^2(Q)$ such that*

$$-\Delta \psi - 2\zeta \cdot \nabla \psi = f. \tag{6.8}$$

Moreover we have

$$\|\psi\|_{L^2(Q)} \leq \frac{R}{\pi |s|}\|f\|_{L^2(Q)}.$$

Proof. Using the Hilbert basis $(\varphi_\alpha)_{\alpha \in Z_0}$ defined in (6.6), one sees that equation (6.8) is equivalent to:

$$\forall \alpha \in Z_0, \quad (-\Delta \psi - 2\zeta \cdot \nabla \psi | \varphi_\alpha) = (f | \varphi_\alpha),$$

where $(\cdot\,|\,\cdot)$ denotes the scalar product of $L^2(Q)$. Using the above remarks summarized in (6.7), we have

$$(-\Delta\psi - 2\zeta\cdot\nabla\psi\,|\,\varphi_\alpha) = (\alpha\cdot\alpha - 2i\zeta.\alpha)\,(\psi\,|\,\varphi_\alpha).$$

It follows that, provided $\alpha\cdot\alpha - 2i\zeta\cdot\alpha \neq 0$, we can express $(\psi\,|\,\varphi_\alpha)$ in terms of $(f\,|\,\varphi_\alpha)$. However this is the case, since $i\zeta\cdot\alpha = is\alpha_1 - \xi\cdot\alpha$ and therefore

$$|\alpha\cdot\alpha - 2i\zeta\cdot\alpha| \geq |\Im\mathrm{m}(\alpha\cdot\alpha - 2i\zeta\cdot\alpha)| = |2s\alpha_1| \geq \frac{|s|\pi}{R}.$$

Therefore we have $(\psi\,|\,\varphi_\alpha) = (f\,|\,\varphi_\alpha)/(\alpha\cdot\alpha - 2i\zeta\cdot\alpha)$ and

$$\psi = \sum_{\alpha\in Z_0} \frac{(f\,|\,\varphi_\alpha)}{\alpha\cdot\alpha - 2i\zeta\cdot\alpha}\,\varphi_\alpha\,.$$

This, since $(\varphi_\alpha)_\alpha$ is a Hilbert basis of $L^2(Q)$, yields the estimate

$$\|\psi\|^2_{L^2(Q)} = \sum_{\alpha\in Z_0} \frac{|(f\,|\,\varphi_\alpha)|^2}{|\alpha\cdot\alpha - 2i\zeta\cdot\alpha|^2} \leq \frac{R^2}{|s|^2\pi^2}\|f\|^2_{L^2(Q)},$$

and, using the regularity results for elliptic equations, we have also $\psi \in H^2(Q)$. ☐

Corollary 6.4. *Let* $\zeta := i\xi + \eta$ *where* $\xi, \eta \in \mathbb{R}^N$ *are such that* $\xi\cdot\eta = 0$ *and* $\eta \neq 0$. *Then if* $\Omega \subset \mathbb{R}^N$ *with* $N \geq 2$ *is a bounded domain, there exists a linear continuous operator* $B_\zeta : L^2(\Omega) \longrightarrow H^2(\Omega)$ *such that for any* $f \in L^2(\Omega)$, *the function* $\psi := B_\zeta(f)$ *satisfies*

$$\psi \in H^2(\Omega), \quad -\Delta\psi - 2\zeta\cdot\nabla\psi = f. \tag{6.9}$$

Moreover for some constant $c > 0$, *independent of* ξ, η *and* f, *we have*

$$\|B_\zeta(f)\|_2 \leq \frac{c}{|\eta|}\,\|f\|_2\,.$$

Proof. Without loss of generality we may assume that $0 \in \Omega$, and we may fix $R > 0$ such that for any rotation S of $\mathbb{R}^N$ around the origin we have $S(\Omega) \subset (-R, R)^N$. For a given $\eta \in \mathbb{R}^N \setminus \{0\}$, we may set $s := |\eta|$ and find a rotation S in $\mathbb{R}^N$ such that $Se_1 = \eta/s$. If $f : \Omega \longrightarrow \mathbb{C}$ is a function in $L^2(\Omega)$, first we may extend it to all the space $\mathbb{R}^N$ by setting $\widetilde{f}(x) := 0$ whenever $x \notin \Omega$ and $\widetilde{f}(x) = f(x)$ for $x \in \Omega$. Next one checks that if $u : \mathbb{R}^N \longrightarrow \mathbb{R}$ is a function in $H^1_{\mathrm{loc}}(\mathbb{R}^N)$, upon setting $v(x) := u(Sx)$ one has $\nabla v(x) = S^*(\nabla u)(Sx)$ and

$$\zeta_0\cdot\nabla v(x) = (S\zeta_0)\cdot(\nabla u)(Sx), \quad \Delta v(x) = \Delta u(Sx).$$

It follows that if we set $\zeta_0 := iS^*\xi + S^*\eta$, then in order to solve (6.9) in Ω it is enough to set $g(x) := \widetilde{f}(Sx)$ and solve

$$v \in H^2(Q) \cap H^1_{Z_0}(Q), \quad -\Delta v - 2\zeta_0\cdot\nabla v = g.$$

Since this has been done in Proposition 6.3 with $\|v\|_{L^2(Q)} \le c\|g\|_{L^2(Q)}/|\eta|$, one sees that if we set $B_\zeta(f) := v_{|\Omega}$, all the requirements of the corollary are satisfied.   □

Now we can solve equation (6.3) provided $|\zeta|$ is large enough.

Proposition 6.5. *Let* $\zeta := i\xi + \eta$ *where* $\xi, \eta \in \mathbb{R}^N$ *are such that* $\xi \cdot \eta = 0$ *and* $\eta \ne 0$. *Then if* $\Omega \subset \mathbb{R}^N$ *(with* $N \ge 2$*) is a bounded domain and* $q \in L^\infty(\Omega)$*, there exist a constant* $c_0 > 0$ *and for* $|\eta| \ge c_0$ *a linear continuous operator* $M_\zeta : L^2(\Omega) \longrightarrow H^2(\Omega)$ *such that for any* $f \in L^2(\Omega)$*, the function* $\psi := M_\zeta(f)$ *satisfies*

$$\psi \in H^2(\Omega), \quad -\Delta\psi - 2\zeta \cdot \nabla\psi + q\psi = f. \tag{6.10}$$

Moreover for some constant $c > 0$*, independent of* ξ, η *and* f*, we have*

$$\|\psi\|_2 = \|M_\zeta(f)\|_2 \le \frac{c}{|\eta|} \|f\|_2 .$$

Proof. The operator B_ζ being as in Corollary 6.4, we note that solving (6.10) is equivalent to solve $-\Delta\psi - 2\zeta \cdot \nabla\psi = f - q\psi$ and hence

$$\psi = B_\zeta(f) - B_\zeta(q\psi). \tag{6.11}$$

If we show that for $|\eta|$ large enough $\psi \mapsto B_\zeta(q\psi)$ is a contraction from $L^2(\Omega)$ into itself, then the mapping $\psi \mapsto B_\zeta(f) - B_\zeta(q\psi)$ has a unique fixed point and equation (6.11), and thus (6.10), has a unique solution.

But $\|B_\zeta(q\psi)\|_2 \le c|\eta|^{-1}\|q\psi\|_2 \le c|\eta|^{-1}\|q\|_\infty\|\psi\|_2$, and therefore when, for instance $|\eta| \ge 2c\|q\|_\infty$, we have $\|B_\zeta(q\psi)\|_2 \le 2^{-1}\|\psi\|_2$ and thus equation (6.11) has a unique solution. Also if we set $A_\zeta(\psi) := B_\zeta(q\psi)$ we have $\|A_\zeta\| \le 1/2$ for $|\eta| \ge 2c\|q\|_\infty$ and the solution ψ of (6.11) is given by $\psi = (I + A_\zeta)^{-1}B_\zeta(f)$. Whence if we set $M_\zeta := (I + A_\zeta)^{-1}B_\zeta$, we have

$$\|\psi\|_2 = \|M_\zeta(f)\|_2 \le \|(I + A_\zeta)^{-1}\| \, \|B_\zeta(f)\|_2 \le \frac{2c}{|\eta|} \|f\|_2 ,$$

since $\|(I + A_\zeta)^{-1}\| \le 2$. Finally using classical regularity results in equation (6.10) one sees that $\psi \in H^2(\Omega)$, and the proposition is proved.   □

We are now in a position to prove the result announced in Theorem 5.7.

Theorem 6.6. *Theorem. Assume that* $\Omega \subset \mathbb{R}^N$ *is a bounded domain and that* $N \ge 3$. *For* $q_j \in L^\infty(\Omega)$ *(with* $j = 0, 1$*) let* $V_0(q_0, q_1)$ *be the space*

$$V_0(q_0, q_1) := \left\{ u_0 u_1 : u_j \in H^1(\Omega), \text{ and } -\Delta u_j + q_j u_j = 0 \right\}.$$

Then $V_0(q_0, q_1)$ *is dense in* $L^1(\Omega)$.

Proof. Let $\xi \in \mathbb{R}^N$ with $\xi \ne 0$ be given. It is interesting to emphasize that the assumption $N \ge 3$ comes in at this point: when the dimension $N \ge 3$, there exist at least two other directions η, σ orthogonal to ξ and such that η is orthogonal to σ. That is we may fix η and $\sigma \in \mathbb{R}^N$ such that for a given $t > 0$ we have:

$$\eta \cdot \xi = \xi \cdot \sigma = 0, \quad |\sigma| = 1, \quad |\eta|^2 = |\xi + t\sigma|^2 = |\xi|^2 + t^2.$$

Then consider

$$\zeta_0 := \eta + i(\xi + t\sigma), \quad \zeta_1 := -\eta + i(\xi - t\sigma),$$

so that $x \mapsto \exp(\zeta_j \cdot x)$ is harmonic in $\mathbb{R}^N$, and let ψ_j be the solution given by Proposition 6.5 solving (6.3). Or rather more precisely let ψ_j satisfy:

$$\psi_j \in H^2(\Omega), \quad -\Delta \psi_j - 2\zeta_j \cdot \nabla \psi_j + q_j \psi_j = -q_j,$$

in such a way that $u_j := (1 + \psi_j) \exp(\zeta_j \cdot x)$ solves

$$u_j \in H^2(\Omega), \quad -\Delta u_j + q_j u_j = 0,$$

and $u_0 u_1 \in V_0(q_0, q_1)$. In order to show that $V_0(q_0, q_1)$ is dense in $L^1(\Omega)$, we have to prove that if for some $h \in L^\infty(\Omega)$ and for all $w \in V_0(q_0, q_1)$ we have $\int_\Omega h(x) w(x) \, dx = 0$, then necessarily $h = 0$. Now choosing $w := u_0 u_1$, we have

$$\int_\Omega h(x) u_0(x) u_1(x) \, dx = \int_\Omega h(x)(1 + \psi_0(x))(1 + \psi_1(x)) \exp((\zeta_0 + \zeta_1) \cdot x) \, dx = 0,$$

and since $\zeta_0 + \zeta_1 = 2i\xi$, we conclude that

$$\int_\Omega h(x) e^{2i\xi \cdot x} \, dx = -\int_\Omega h(x) e^{2i\xi \cdot x} \left(\psi_0(x) + \psi_1(x) + \psi_0(x)\psi_1(x) \right) \, dx.$$

Therefore

$$\left| \int_\Omega h(x) e^{2i\xi \cdot x} \, dx \right| \leq \|h\|_\infty (\|\psi_0\|_2 + \|\psi_1\|_2 + \|\psi_0\|_2 \|\psi_1\|_2),$$

and using the estimates $\|\psi_j\|_2 \leq c|\eta|^{-1}$ and the fact that $|\eta| \geq t$ we conclude, as $t \to +\infty$, that $\int_\Omega h(x) e^{2i\xi \cdot x} \, dx = 0$, and hence $h = 0$. $\square$

References

[1]   R. A. Adams, *Sobolev Spaces*, Academic Press, New York 1975.

[2]   V. Adolfsson and L. Escauriaza, $C^{1,\alpha}$ domains and unique continuation at the boundary, *Comm. Pure Appl. Math.* 50 (1997), 935–969.

[3]   G. Alessandrini, Stable determination of conductivity by boundary measurements, *Appl. Anal.* 27 (1988), 153–172.

[4]   G. Alessandrini, Singular solutions of elliptic equations and the determination of conductivity by boundary measurements, *J. Differential Equations* 84 (1990), 252–273.

[5]   G. Alessandrini, Stable determination of a crack conductivity from boundary measurements, *Proc. Roy. Soc. Edinburgh Sect. A* 123 (1993), 497–516.

[6]   G. Borg, Eine Umkehrung der Sturm-Liouville Eigenwertaufgabe, *Acta Math.* 78 (1946), 1–96.

[7]   H. Brezis, *Analyse Fonctionnelle, Théorie et Applications*, Editions Masson, Paris 1983.

[8]   R. M. Brown, Global uniqueness in the impedance imaging problem for less regular conductivities, *SIAM J. Math. Anal.* 27 (1996), 1049–1056.

[9]   R. M. Brown and G. Uhlmann, Uniqueness in the inverse conductivity problem for nonsmooth conductivities in two dimensions, *Comm. Partial Differential Equations* 22 (1997), 1009–1027.

[10] A. P. Calderón, On an inverse boundary value problem, in *Seminars on Numerical Analysis and its Application to Continuum Physics*, Soc. Brasil. Mat., Rio de Janeiro, 1980, 65–73.

[11] B. Canuto and O. Kavian, Determining coefficients in a class of heat equations via boundary measurements, *SIAM J. Math. Anal.* 32 (2001), 963–986.

[12] B. Canuto and O. Kavian, Determining two coefficients in elliptic operators via boundary spectral data: a uniqueness result, to appear in *Boll. Un. Mat. Ital.*.

[13] S. Chanillo, A problem in electrical prospection and an n-dimensional Borg-Levinson theorem, *Proc. Amer. Math. Soc.* 108 (1990), 761–767.

[14] R. Courant and D. Hilbert, *Methods of Mathematical Physics*, Wiley, New York 1953 (Volume 1), 1962 (Volume 2).

[15] N. Garofalo and F.H. Lin, Unique continuation for elliptic operators: a geometric-variational approach, *Comm. Pure Appl. Math.* 40 (1987), 347–366.

[16] I. M. Gel'fand and B. M. Levitan, On determination of a differential equation from its spectral function, *Izv. Akad. Nauk. SSSR Ser. Mat.* 15 (1951), 309–360; *Amer. Math. Soc. Transl. Ser.* (2) 1 (1955), 253–304.

[17] D. Gilbarg and N. S. Trudinger, *Elliptic Partial Differential Equations of Second Order*, Grundlehren Math. Wiss. 224, Springer-Verlag, Heidelberg 1983.

[18] P. Hähner, A periodic Faddeev-type solution operator, *J. Differential Equations* 128 (1996), 300–308.

[19] V. Isakov, On uniqueness of recovery of a discontinuous conductivity coefficient, *Comm. Pure Appl. Math.* 41 (1988), 865–877.

[20] V. Isakov, Uniqueness and stability in multidimensional inverse problems, *Inverse Problems* 9 (1993), 579–621.

[21] V. Isakov, *Inverse Problems for Partial Differential Equations*, Appl. Math. Sci. 127, Springer-Verlag, New York 1998.

[22] O. Kavian, Remarks on the determination of coefficients in elliptic operators via boundary measurements, in preparation.

[23] O. Kavian, *Introduction à la Théorie des Points Critiques et Applications aux Problèmes Elliptiques*, Math. Appl. 13, Springer-Verlag, Paris 1993.

[24] A. Kirsch, *An Introduction to the Mathematical Theory of Inverse Problems*, Appl. Math. Sci. 120, Springer-Verlag, New York 1996.

[25] R. V. Kohn and M. Vogelius, Determining conductivity by boundary measurements, *Comm. Pure Appl. Math.* 37 (1984), 289–297.

[26] R. V. Kohn and M. Vogelius, Determining conductivity by boundary measurements, interior results, II, *Comm. Pure Appl. Math.* 38 (1985), 643–667.

[27] O. A. Ladyženskaja and N. N. Ural'ceva, *Equations aux Dérivées Partielles de type Elliptique*, Editions Dunod, Paris 1968.

[28] N. Levinson, The inverse Sturm-Liouville problem, *Mat. Tidsskr. B* (1949), 25–30.

[29] J. L. Lions, *Problèmes aux Limites dans les Équations aux Dérivées Partielles*, Séminaires de Mathématiques Supérieures, No. 1 (Été 1962), Presses de l'Université de Montréal, Montréal 1965.

[30] J. L. Lions and E. Magenes, *Problèmes aux Limites Non Homogènes et Applications (volume 1)*, Editions Dunod, Paris 1968.

[31] V. G. Maz'ja, *Sobolev Spaces*, Springer-Verlag, Berlin 1985.

[32] A. I. Nachman, Reconstructions from boundary measurements, *Ann. of Math.* 128 (1988), 531–577.

[33] A. I. Nachman, Global uniqueness for a two-dimensional inverse boundary value problem, *Ann. of Math.* 142 (1996), 71–96.

[34] A. I. Nachman, J. Sylvester and G. Uhlmann, An n-dimensional Borg-Levinson theorem, *Comm. Math. Phys.* 115 (1988), 595–605.

[35] F. Natterer, *The Mathematics of Computerized Tomography*, John Wiley & Sons and B.G. Teubner, Stuttgart, 1989.

[36] J. Pöschel and E. Trubowitz, *Inverse Spectral Theory.* Academic Press, London 1987.

[37] G. Stampacchia, *Équations Elliptiques du Second Ordre à Coefficients Discontinus*, Séminaires de Mathématiques Supérieures, No. 16 (Été, 1965), Presses de l'Université de Montréal, Montréal 1965.

[38] J. Sylvester and G. Uhlmann, Global uniqueness theorem for an inverse boundary problem, *Ann. of Math.* 125 (1987), 153–169.

[39] J. Sylvester and G. Uhlmann, *The Dirichlet to Neumann map and applications,* in *Inverse Problems in Partial Differential Equations* (D. Colton, R. Ewing and W. Rundell, eds.), SIAM Publications, Philadelphia, 1990, 101–139.

[40] G. Uhlmann, Inverse boundary value problems and applications, *Astérisque* 207 (1992), 153–211.

[41] K. Yosida, *Functional Analysis*, Grundlehren Math. Wiss. 123, Springer-Verlag, New York 1974.

www.ingramcontent.com/pod-product-compliance
Lightning Source LLC
Chambersburg PA
CBHW081107220326
41598CB00038B/7262